Cosmic Heritage

Peter Shaver

Cosmic Heritage

Evolution from the Big Bang to Conscious Life

 Springer

Dr. Peter Shaver
Sydney and Toronto
p4shaver@gmail.com

ISBN 978-3-642-20260-5   e-ISBN 978-3-642-20261-2
DOI 10.1007/978-3-642-20261-2
Springer Heidelberg Dordrecht London New York

Library of Congress Control Number: 2011932248

Cover image: Vincent van Gogh. The Starry Night. Saint Rémy, June 1889.
Oil on canvas, 29 × 36 1/4″ (73.7 × 92.1 cm). Acquired through the Lillie P. Bliss Bequest.
The Museum of Modern Art, New York, NY, U.S.A.
Photo Credit: Digital Image © The Museum of Modern Art/Licensed by SCALA/Art
Resource, NY.

Cover design: eStudio Calamar S.L.

Printed on acid-free paper

Springer is part of Springer Science+Business Media (www.springer.com)

To Jenny, Nikki and Adam

Contents

1. Setting the Scene

Our connections to the early universe are profound. The universe and its contents have evolved continuously all the way from the Big Bang to the present, and this has made us what we are today. We are part of the universe. This is our Cosmic Heritage.

The very matter we're made of comes from the earliest moments of the universe. The physical laws that govern our universe were there from the start. At one stage darkness turned into light, as stars and galaxies formed. When we now look out into the universe we look back into the past, so we can readily follow the history of the universe by observing galaxies – beads on the string of time.

The elements are constantly being brewed up in stars, and have accumulated over the aeons. The continuing process of star formation led to by-products such as planets, many of which may be suitable habitats for life. Organic molecules formed in the surrounding space. The scene became primed for life.

Our Earth is one of those planets, and life emerged a relatively short time after the Earth was formed. Over the history of the planet a great many species have taken their turn. More than 99% of them eventually became extinct, but they are always being replaced by newly-evolved species.

We have come to realize that all living things on Earth, including ourselves, are members of one single family. And that life itself is just based on 'information'. This information is the code of life, common to all life forms, from bacteria to us. It is written and saved in our genomes. The atoms and molecules of which we are made may come and go, but the information written in our genomes remains with us forever.

Brains have evolved as much as anything else in our bodies, and our brains happen to have become exceptional. As a result, at the moment we humans are dominant on this planet, and

P. Shaver, *Cosmic Heritage*, DOI 10.1007/978-3-642-20261-2_1,

undoubtedly unique in being able to contemplate the distant universe.

We know that planets exist around other stars – perhaps billions – and some may also harbour living creatures. We have no idea how these may compare with us. Given the huge time-scales available in our universe, they would almost certainly be millions or billions of years more advanced or less advanced than we are. In any case, we are probably not alone in the universe.

This can all be put into perspective by compressing the entire 14-billion year history of the universe into just 1 year. The Big Bang occurred with great fireworks at the very start of the new year. The first stars and galaxies had emerged by mid-January, although our own Sun and Earth didn't form until early September. Later that month the first primordial life appeared on Earth. But it was only in December that complex life appeared and the evolution of life on Earth really took off, and it wasn't until 30 December that the famous extinction of the dinosaurs took place. Recorded human history started just seconds before midnight on New Year's Eve, and modern technology in the last fraction of a second.

It is impressive that scientists have been able to piece this story together, from such diverse fields of research and with such rigour. But of course many mysteries remain, and we have no idea how much further the story will take us. We can imagine, but we don't know what the future may hold.

To set the scene for our cosmic story, this first chapter provides a very brief tour of the universe. The next few chapters then provide essential background leading up to the chapter on the Big Bang, and thereafter the book follows the evolution of the universe and life to the present and beyond.

What's Out There?

A glance up at the sky at night gives little indication of the drama of the universe. The Moon and planets follow their predictable courses, and the stars appear to sit fixed in their places. There is a deep sense of peace. Only the darkness of the night sky betrays the violence of the distant universe.

In Our Solar System

The solar system is our cosmic backyard. At its centre is the Sun, our local star, and a typical one at that. The Sun, like all normal stars, derives its energy from nuclear fusion; it 'burns' hydrogen into helium. The Sun dwarfs everything else in the solar system, all of which, including the planets, is just debris left over from the formation of the Sun 4.6 billion years ago. The eight planets, of which Earth is one, orbit the Sun in a nearly flat disk, and moons like ours also orbit most of the other planets. Interspersed with the planets are much smaller bodies, rocky asteroids, and icy comets are sometimes swept in from the outer regions of the solar system.

To give an idea of relative scales, let's start with something fairly small and close – our Moon. Its diameter, 3,500 km, is about a quarter that of Earth. Its distance from Earth, 384,000 km, corresponds to only 18 days of commercial flying (about ten round-trip flights from Europe to Australia). A one-way trip to the Moon on Apollo, including various orbits and manoeuvres, took a few days. These are scales we can easily grasp.

The other planets in the solar system range in diameter from about a third that of Earth (Mercury) to 11 times (Jupiter). The mass of Mercury is only 6% that of Earth, and the mass of Jupiter is over 300 times that of Earth. While the Earth takes a year to complete its orbit around the Sun, close-in Mercury whizzes around in a quarter of a year and far-out Neptune takes a leisurely 165 years. The planets are neatly placed in two categories: the small inner rocky planets (Mercury, Venus, Earth and Mars), and the large outer gaseous planets (Jupiter, Saturn, Uranus and Neptune). Six of the eight planets have moons orbiting around them; the three largest planets each have dozens of moons.

Dominating the solar system is the Sun. It accounts for 99.9% of the mass of the solar system. It is over 300,000 times more massive than the Earth, and about a hundred times bigger in diameter.

The distances involved start to become impressive when you consider the whole solar system. Even Earth, one of the inner planets, is some 150 million kilometres from the Sun. This is so far that it takes *light* over eight minutes to travel from the Sun to the

The Sun is eclipsed by Saturn in this spectacular image taken by the Cassini spacecraft in 2006. Our Earth is the tiny dot which may be just visible between the upper left portions of the outer rings of Saturn. At that time Cassini was about 2 million kilometres from Saturn, and the Earth was 1.5 billion kilometres in the background. Makes you appreciate the immensity of the solar system, let alone the universe. Image courtesy of NASA/JPL/Space Science Institute. The Cassini mission is an international collaboration between NASA, ESA and several other partners. Credit: CICLOPS, JPL, ESA, NASA

Earth[1]; seven minutes ago the Sun may have switched off and we wouldn't know it yet (although we shouldn't worry too much about this, as the Sun is expected to last for another several billion years).

The far outer reaches of the solar system are occupied by the Oort Cloud, a spherical zone of billions of icy comets, which extends out to a light-year (about nine trillion kilometres) from the Sun. The total extent of the solar system may be considered to be that over which the Sun's gravitational pull exceeds that of nearby stars, the closest of which is four light-years away. At the remote boundaries of our solar system our Sun would appear no brighter than several other stars visible in the sky, and its gravitational field would have almost faded into the galactic background. A size of trillions of kilometres may seem big, but it's tiny on the scale of the universe.

How do we know so much about distant objects? In the solar system we have had the advantage of spacecraft missions over the last half century. We have sent probes to all the planets and several moons, and landed spacecraft on Mars and our Moon, including several manned missions to the Moon. One spacecraft, NASA's Pioneer 10, has gone well beyond the most distant planets and will eventually leave the solar system entirely. These have all brought us close-up views, copious amounts of detailed information, and even return samples. But the fly-by probes have still relied on passive observations from a distance, as do (obviously) studies of the Sun. And for objects beyond our solar system, passive observation is all we have. How can we know so much from 'mere observation'?

The two main modes of observation are imaging and spectroscopy. Imaging is pretty obvious; we want the sharpest and most sensitive images we can get. In spectroscopy we spread out the spectrum of light into its component colours (wavelengths), from red to blue, using a spectrograph. We're all familiar with the

---

[1]Light takes time to travel. Its speed is 300,000 km/s. The speed of light is constant, and nothing travels faster than light; the speed of light *in vacuo* is the 'ultimate speed limit'. Therefore, astronomical distances are often expressed in 'light-years', the distance that light can travel in one year. That distance is about 9 trillion kilometres, which, written out, is 9,000,000,000,000 km, or simply 9×10^{12} km. (The latter 'scientific' notation will sometimes be used in this book for convenience. The superscript 12 gives the number of zeroes following the 9. A small number can be expressed in a similar way: 0.004 is 4×10^{-3}.)

appearance of a spectrum. The same process gives us the colours of a rainbow, and we can reproduce it easily at home with sunlight shining through a prism. The spectrum of an astronomical object gives us a huge amount of information. Usually we can see sharp, narrow bright (emission) or dark (absorption) features at specific wavelengths in the spectrum. These are due to atoms and molecules in the distant object, and are referred to as emission and absorption lines. There can be anywhere from a few to thousands of these lines in a given spectrum. Their relative strengths tell us the chemical composition of the object. The lines can also be shifted along the spectrum by the motion of the object. If the object is moving towards us, the lines are shifted towards the blue, and if the object is moving away from us, the lines are shifted towards the red. These are called blueshifts and redshifts, and the phenomenon is commonly referred to as the 'Doppler effect'. We experience the acoustic Doppler effect when we hear the siren of a speeding ambulance: it is high-pitched when approaching us, and low-pitched when going away.

The light we normally see (with our own eyes) is referred to in astronomy as 'visible' or 'optical' light. It is actually a narrow part of the whole electromagnetic spectrum, which ranges from short wavelengths (gamma-rays, X-rays, ultraviolet rays) to long wavelengths (infrared, millimetre and radio), with the visible part in the middle. The atmosphere of the Earth is opaque to much of the electromagnetic spectrum, and only the visible and radio wavelengths can easily get through. Therefore, we can only use optical and radio telescopes on the ground; the rest has to be done using satellites or spacecraft above the atmosphere. Almost all our observations are made using the electromagnetic spectrum.

Such observations have made possible sophisticated and detailed knowledge about objects far beyond our solar system, in our galaxy, and even in the very distant universe, as we shall see.

In Our Galaxy

Our solar system resides in a comfortable neighbourhood of an ordinary disk-like spiral galaxy. The nearest stars, the ones we can easily see at night, are four light-years or more away from us, and the full extent of our galaxy is over 80,000 light-years (almost a million trillion kilometres). We are located in the flat disk of this galaxy, which can therefore be seen edge-on – it is the Milky Way band extending across the sky, most prominently visible in the southern hemisphere. The diffuse 'milky' appearance is due to the approximately 100 billion stars crowded into the plane of our galaxy, most of which are similar to our Sun.

Our galaxy contains a wonderful 'zoo' of astronomical phenomena (as do all galaxies, but in our own galaxy we can see them close-up). Almost all of the visible contents of our galaxy are related in one way or another to stars, so a good way to make an inventory of the contents of our galaxy is to follow the life cycles of stars.

All stars form in essentially the same way. They originate in dense clouds of gas and dust. Our galaxy, like all galaxies, contains an interstellar medium. That is, the space between the stars is not completely empty, but rather contains a dilute distribution of atoms, molecules and dust particles spread throughout our galaxy. The average density, about one atom per cubic centimetre, is very low by comparison with the best vacuums we can produce here on Earth, but it is enough to ultimately produce the hundreds of billions of stars in our galaxy. The present interstellar medium is comprised of about 70% hydrogen, 28% helium, and 2% heavier elements by mass.

The interstellar medium is not perfectly uniform. Its density varies, and it is these variations that make the formation of stars possible. All matter attracts other matter through the force of gravity, and any over-dense region will grow and become denser as it sweeps up matter from adjacent regions of the interstellar medium. Over the course of time such concentrations become dense enough that complex molecules and dust grains can form and grow into what we call molecular clouds. These are the nurseries of stars. We can see them with the naked eye by looking at the Milky Way at night; the molecular clouds are opaque and block out the light from background stars, so they appear as black splotches along the Milky Way.

These molecular clouds continue to become denser and denser with time, through gravitational attraction, and when the density becomes great enough, a star is born. This process will be described in some detail in Chap. 8, but here we just point out some of the pyrotechnics produced in the process. While the infalling matter becomes more and more concentrated into a rapidly spinning accretion disk (also known as a protostellar disk or protoplanetary disk), emerging out of the opposite poles of the opaque accretion disk we eventually see bright, narrow, linear 'jets' of emission, which can illuminate matter lying in their paths. These are called protostellar jets, striking features which are hallmarks of young star formation.

When a fully-developed massive young star has been formed, its heat pressure starts to blow most of the parent molecular cloud away, and its bright light is reflected through gaps opening up in the cloud, in what are referred to as reflection nebulae. Much more impressive, though, is when the star has cleared a large volume and its intense radiation ionizes the inner regions of the fragmenting cloud (removing negatively charged electrons from atoms to form ions), which causes these regions to glow, rather like neon lights but on an enormous scale. These are emission nebulae; they produce some of the most spectacular and famous images seen in astronomy, such as the Orion Nebula. Emission nebulae are found wherever massive new stars are being formed; they are so prominent that they serve as beacons to help us find new regions of star formation.

After the star has cleared its surroundings, including the gas and dust of the protostellar disk, some debris still remains, too massive and compact to be blown away. This debris includes planets, their moons and asteroids. These become permanent residents, orbiting around the newly born star. Further out is a vast cloud of icy comets, the last remnants of the original molecular cloud.

Until 1995 the only planets we knew were those in our own solar system, and anything else was speculation. Now we have discovered over 500 planets orbiting other stars, and soon that number will rise into the thousands. It seems that most stars are accompanied by planets; if so, there may be billions of planets in our galaxy, many of which may support life. The fundamental discovery of the first 'extrasolar' planets in the 1990s will be described in detail in Chap. 17.

Once a star has blown away its parent cloud, it lives much of its life in splendid isolation in space, along with its planetary entourage, a sphere of hot gas like the Sun. Its brightness is due to ongoing nuclear fusion processes. We see stars all around us, but as we are located within the disk of a spiral galaxy, which we therefore see edge-on, most of the stars we see are concentrated along the Milky Way, the plane of our galaxy. There are many types of stars, the differences being largely based on mass. They include brown dwarfs ('failed stars' that didn't have enough mass for nuclear burning), normal stars like our Sun that live for billions of years, and massive stars that burn themselves out in just a few million years.

In later life stars go through various convulsions, which cause some of their outer regions to be ejected. We see these as illuminated and ionized shells surrounding the stars. The most famous and spectacular are the planetary nebulae, so called because in early telescopes they appeared to be circular disks of light, similar to the early images of other planets in our solar system. Nowadays we know exactly what they are and what they're made of. The Hubble Space Telescope website contains awe-inspiring images of hundreds of planetary nebulae.

By far the most impressive of the stellar end of life convulsions are the gigantic explosions called supernovae. One supernova explosion can be as bright as an entire galaxy of 100 billion stars. They typically occur in an average galaxy once every century or so. In our own galaxy, in spite of the obscuring dust in the edge-on plane that we observe, four supernovae have been observed and recorded over the past millennium, so we know their locations and ages. Luckily, none of them was too close. Supernovae reach their peak luminosities quickly, and are still as bright as billions of stars for a week or so. The vast shell of ejected material is called a supernova remnant. It can shine brightly at some wavelengths for a hundred thousand years or so, until it just merges with the general interstellar medium.

Stars produce and distribute the 'heavy elements' such as carbon, which are vital for life as we know it. The convulsions that stars undergo late in their lives, including supernova explosions, are an essential part of this process.

What remnant is left at the end of a star's lifetime? Again, it depends on the initial mass of the star. In the case of a low-mass or typical star like our Sun, the remnant becomes a white dwarf star, so called because it is hot and small. From that point on it just gradually becomes cooler and less luminous.

The core of a star whose original mass was more than eight solar masses collapses to become a neutron star, comprised largely of closely-packed subatomic particles called neutrons. A neutron star is incredibly dense – about as dense as an atomic nucleus. It is typically just 10 km in radius, yet more massive than the Sun. A teaspoon of its material would have a mass of more than a trillion kilograms. As it formed from a star that was spinning, albeit slowly, it ends up spinning rapidly (like a figure skater pulling in her arms), with periods ranging from seconds to a thousandth of a second, and almost exactly at a constant rate. It can be as good a timer as an atomic clock.

If neutron stars do indeed rotate with typical periods of the order of a second, and if they were to produce narrow beams of emission, pulses could be observed each time the beam pointed towards the Earth, just as a rotating lighthouse beam is seen as a series of flashes. This is just what is observed, and the objects are called pulsars.

Black holes are even more famous. The core of a star whose original mass was over 25 solar masses will keep collapsing without end, ultimately becoming a black hole. This is absolutely unavoidable. Unlike white dwarfs and neutron stars, which are blocked from collapsing beyond certain points by the fundamental laws of physics, nothing can stop the collapse of a sufficiently massive star. It collapses 'all the way', and becomes a black hole.

Nothing can escape from a black hole, not even light. A black hole is caused by the extreme deformation of space by a very compact mass; it is the ultimate space-time warp. The boundary around a black hole at which the speed needed to escape the gravitational attraction of the black hole equals the speed of light is called the event horizon. There is no way we can know anything about what happens inside the event horizon.

Although they are themselves invisible, black holes can still be detected through their effects and interactions with other matter. Black holes can cause the bending of light from distant objects behind them, and close interactions can result in matter spiralling into a black hole, generating great heat and light. Black holes would seem to mark a definitive end to the life cycle of stars. But do they?

Just because a star is dead doesn't mean it can't be resurrected. Many stars are found in binary systems, containing two stars orbiting around each other. When one member of a close binary pair dies (for example as a white dwarf) the other can bring it back to life. When the companion reaches the 'convulsive' stage in its life, its loosely held outer gas can be transferred onto the white dwarf, providing it with a new energy source. As the matter accumulates on and around the white dwarf it gets hotter, ultimately reaching the point at which nuclear fusion can begin. This causes a thermonuclear flash on the surface of the white dwarf, and the binary system attains the luminosity of a hundred thousand stars for a period of a few weeks. This is called a nova. The accreted material is ejected, and the accumulation process starts all over again.

Binaries involving neutron stars are similar in principle, but the gravitational fields and energies are much greater. The gas accreting around the neutron star is so hot that it emits copiously in energetic X-rays. For this reason these are called X-ray binaries. The thermonuclear bursts in these cases are short (a few seconds), but they radiate a hundred thousand times the luminosity of the Sun. They recur over periods of hours to a few days. Some X-ray binaries may contain black holes rather than neutron stars; the most convincing case is an object called Cygnus X-1, which contains a star 18 times more massive than the Sun orbiting an unseen companion which, from X-ray spectroscopy, is almost certainly much more massive than a neutron star.

There is one important phenomenon in our galaxy which has nothing to do with the standard stellar life cycle: the galactic centre. One might easily suspect that something special must be happening at the very centre of our galaxy, but what? The galactic centre is totally obscured from our view by intervening interstellar dust, as both we and it are located in the relatively dense plane of the galaxy, and the distance between us and the galactic centre is large: 27,000 light-years, or 260 million billion kilometres. However, complete obscuration only occurs at optical wavelengths; in other regions of the electromagnetic spectrum (such as the radio, millimetre, infrared and X-ray bands), the view is essentially unobscured. Through meticulous observations we now know that there is a supermassive black hole at the very centre of our galaxy, with a mass 4 million times that of our Sun.

So there you have it – the galactic zoo, which includes molecular clouds, protostars, protostellar disks and jets, reflection nebulae, emission nebulae, planets, the variety of stars, planetary nebulae, novae, supernovae, brown dwarfs, white dwarfs, neutron stars, pulsars, X-ray binaries and black holes. And the galactic centre with its supermassive black hole.

Beyond Our Galaxy

A hundred years ago it was thought that our galaxy was the entire universe. Now we know that the universe is enormously bigger. Beyond our galaxy lie vastly more galaxies – many billions of them, each containing anywhere from tens of millions to trillions of stars like our Sun. Our galaxy is just average. Typical galaxies are tens to hundreds of thousands of light-years in size. They are separated from each other by millions of light-years, and the density of matter in the space between them is of the order of one atom per cubic metre. The galaxies are 'dots' in a universe that is billions of light-years in size.

The same objects and phenomena that we described in our galaxy are commonplace in the billions of other normal galaxies of various types spread throughout the universe. There are minor differences – ours is a spiral galaxy, elliptical galaxies have less of an interstellar medium, and irregular galaxies generally have more – but these details don't change the big picture.

By using our most powerful telescopes and the technique of spectroscopy, it has been possible to determine that distant galaxies are made of exactly the same elements and atoms as we find in our own galaxy, and that the same laws of physics apply in the distant universe as here on Earth.

We can observe supernovae exploding in both nearby and distant galaxies. Even brighter events are sometimes observed, probably caused by imploding massive stars or mergers of binary neutron stars. These are called gamma-ray bursts; some of these are the brightest explosions ever observed in the universe (one was visible with the naked eye even though it was 8 billion light-years away), and some are amongst the most distant known objects.

Most galaxies in the local universe are quiescent grand-design ellipticals or spirals, but some galaxies exhibit a wondrous range of exotic behaviours. Pairs of galaxies may be seen doing a sort of 'cosmic dance', rotating closely about each other with arms 'joined'. Others are colliding, essentially passing through each other again and again until they finally merge into one (much larger) galaxy. In some, the supermassive black hole at the centres are being 'fed' by gas and stars in unstable orbits, and spectacular outflows can result. These can cause huge jets and 'bubbles' penetrating the intergalactic medium on opposite sides of the galaxy, and extending far beyond the dimensions of the galaxy itself. These often emit copiously at radio wavelengths, in which case they are called 'radio galaxies'. Even more dramatic are the 'quasars', which we see when the rotation axes of the galaxies are pointed almost straight at us. The light from jets of matter emitted from the nuclear region around the central supermassive black hole is enhanced by an effect of relativity, and all we can normally see of the galaxy is just this brilliant point of light, which appears far brighter than the rest of the galaxy combined. Because of their enormous brightness, quasars can be seen out to vast distances and early times – the light we see from the earliest quasars has been travelling to us for over 13 billion years. It is thought that most if not all galaxies contain black holes at their centres, formed billions of years ago when the galaxies themselves were being formed. In most galaxies today (like ours), the black holes are quiescent because of lack of fuel – they are 'starved monsters'.

Galaxies are not distributed uniformly. They tend to be clustered in groups, and distributed in gigantic filaments and sheets throughout the universe. These are the largest structures known, extending over hundreds of millions of light-years (several billion trillion kilometres). They grew, over the history of the universe, from small primordial fluctuations in the distribution of matter in the very early universe. Their evolution can be traced and their structures replicated extremely well by large computer simulations.

The huge masses of galaxies and clusters of galaxies can actually distort the images we see of more distant galaxies. This results from the fact that gravity can bend light, an effect predicted by Einstein in 1915. The effect is called 'gravitational lensing'. If a massive galaxy is very close to our line of sight to a much more distant galaxy, we can sometimes see two or more images of the distant galaxy; in cases of almost perfect alignment the distant galaxy is smeared out into a ring surrounding the image of the intervening galaxy. When the intervening object is a dense cluster of galaxies, we can see many arcs centred on the cluster. It is a spectacular effect.

Observations by the Hubble Space Telescope (HST), in particular the famous 'Hubble Deep Field', have revealed astonishing views of the distant universe. The Hubble Deep Field is a small region of dark sky, chosen because there happened to be no bright stars or galaxies in that particular direction. It is truly an uncluttered view of the distant universe. It was dubbed a 'blank field'. In 1995 the HST continuously stared at that blank piece of sky for 10 complete days, so the sensitivity reached was phenomenal.

Astronomers were absolutely staggered by the resulting image. It was unlike anything they had ever seen before. The Hubble Deep Field image is dominated by thousands of small, faint, ill-formed galaxies of irregular shape, as far away as the most distant quasars. We are looking out to the distant universe as it was less than a billion years after the Big Bang. At the limit of the most sensitive HST surveys today, we can see a hundred billion galaxies over the whole sky, and there are more. These observations changed our view of the universe forever.

The Frenzied Sky

Everything in the universe is moving. This includes the 'fixed stars' we see with the naked eye, and even the most distant galaxies. They just appear to us to be stationary because, even if their true motions are large, their distances from us are so huge that nothing seems to change (on our timescale). With large telescopes and precision satellites we can now readily measure the motions of stars in our own galaxy.

Everything changes in brightness too, on one timescale or another. Pulsars pulse. Stars have hiccups and sometimes eject huge shells of matter. Distant quasars fluctuate in brightness, sometimes in violent bursts. Novae recur. Supernovae happen only once, but when they do it's impressive, as they can be as bright as an entire galaxy. And these explosions occur across the universe. As there is a supernova event roughly once per century in a typical galaxy (probably more in the early universe of rapid star and galaxy formation), this means that hundreds of thousands or millions of supernovae are going off in our observable universe every day. Look above you and think of the whole sky peppered with millions of outbursts happening all the time. The peaceful night sky mentioned at the beginning of this chapter is actually a frenzy of activity, but most of it is far too faint to be seen with the naked eye.

The Darkness Beyond

The rich zoo of cosmic inhabitants described above does not extend forever. Imagine the following thought experiment, in which we 'peel away' the layers of the universe, one by one, from the nearest to the furthest from us. We start with the familiar sky as seen with the naked eye. The closest objects we see are our neighbours in the solar system – the Sun, the Moon, and the brightest planets. Now imagine that these are 'switched off' – they are no longer visible to us. The sky we see now is dominated by the nearest stars and the familiar patterns of the constellations. Switch these off. The sky will now be dominated by more distant stars in our galaxy, spread over the sky but concentrated towards the diffuse band of the Milky Way, itself comprised of billions of stars. Switch all these off and we turn off our entire galaxy. The sky is now dominated by the nearest galaxies – the Magellanic Clouds, the Andromeda Nebula, and several others. Switch these off. The sky now appears almost uniformly sprinkled with billions of distant galaxies. Finally, we switch all of these off, right out to the distance of the first galaxies and stars. The sky is now totally black. We have reached back to what is sometimes called the 'edge of the universe'. It is actually the near side of the 'cosmic dark ages'. At the far side is the Big Bang and all the activity of the very early universe. These will be discussed in the following chapters.

The diverse objects and phenomena that we see in the universe, and the huge scales of distance and time, may seem astonishing to us. Certainly the cosmos is magnificent and awe-inspiring. But to astronomers the objects in the universe are as real as the distant mountains we see here on Earth. The Apollo Moon landings brought this home to all of us. In some ways astronomical discovery is not so different from the days of geographical exploration hundreds of years ago, when distant shores were not known. We are discovering what is really out there. The distances are impressive, but no more so than, say, the incredibly small scales that we explore in subatomic physics, or the astonishing degree of the complexity of life all around us here on Earth. And we have become accustomed to timescales approaching those in astronomy through the science of geology. We also shouldn't forget that we humans now number almost 7 billion. (That means that the hundred billion stars in our galaxy amount to only about 14 stars per person.) We are ourselves part of the universe of large numbers. Still, the universe is a pretty big place.

Now that we are familiar with many of the inhabitants of the universe, we can move on to consider the large scale properties of the universe as a whole. We enter the field of cosmology.

2. The Universe Is Expanding and Evolving

We now know that the entire universe is both expanding and evolving. These two remarkable facts have been established beyond doubt over the past century.

The simple observation that the sky is dark at night indicates that the universe cannot be infinite and unchanging, comprised of an infinite number of unchanging stars. If that were so, then the stars (which have finite sizes) would overlap in every line of sight, and the sky would be as bright as the surface of the Sun in all directions. This is known as 'Olbers' Paradox'. The darkness of the night sky rules out an unchanging universe that is infinite in space and time.

In astronomy we directly observe the distant past, and can therefore directly study the evolution of the universe. This is possible due to the finite speed of light. It takes time for light to travel from one place to another, from a distant galaxy to us. Therefore when we look out into the distant universe, we also look back in time. The most distant galaxies we see are now being observed as they were over 13 billion years ago – their light has taken that long to reach us.

As mentioned above, the distant universe looks very different from the nearby universe. In contrast to the familiar nearby spiral and elliptical galaxies, the faint distant galaxies have irregular forms beyond imagination. It was a totally different world. Smaller and much more chaotic in appearance, they were young galaxies still in the process of formation. We now know from spectroscopy that these faint galaxies are indeed the most distant (see below), so the early universe was certainly very different from the nearby universe of bright galaxies. From straightforward observations such as these it is already clear that the universe has significantly evolved.

P. Shaver, *Cosmic Heritage*, DOI 10.1007/978-3-642-20261-2_2,
© Springer-Verlag Berlin Heidelberg 2011

The Expansion of the Universe

In 1917 Albert Einstein applied his new equations of general relativity to the universe as a whole. He found, to his discomfort, that they implied that the universe had to be changing with time – either expanding or contracting. The orthodox view at that time was that our galaxy was the entire universe, and the small motions of the stars indicated that it was neither expanding nor contracting. Einstein therefore felt that he had to somehow make his equations consistent with a static universe. He did so by adding what nowadays might be called a 'fudge factor' to the equations – a constant, which became known as the 'cosmological constant'. With this, his equations did indeed produce a static universe, to his satisfaction.

A few years later a young American astronomer by the name of Edwin Hubble started to work at Mount Wilson Observatory in California, using the 100-inch telescope, the largest in the world at that time. He set out to study the mysterious spiral nebulae. There was much debate at the time as to whether these were within our galaxy or outside. He was able to examine what appeared to be individual stars in the Andromeda Nebula, and found that some of them were varying in brightness with a fixed period, similar to the stars known as Cepheid variables in our galaxy. It was already known that there is a relationship between the periods and the luminosities of Cepheids; if you measured the period (the time between peaks in brightness, using photographs taken over several months), you knew the luminosity. And knowing both the intrinsic luminosity and the measured apparent brightness of the star (i.e. the brightness as measured here on Earth), you could determine its distance using the inverse square law (a star of a given luminosity appears four times fainter if it is moved two times further away). In this way Hubble was able to determine the distances to several nebulae. It was clear that they were outside of our galaxy – they were themselves distant galaxies. That was in itself a huge discovery.

But it was only part of the story. Hubble, Georges Lemaître and others also knew, from spectroscopic measurements of the nebulae, how fast these galaxies were moving away from (or towards) us. By 1929 it was clear that almost all of the galaxies are moving away from us, and that the speed of recession (the redshift) increases with the distance to the galaxy. This became known as Hubble's Law. The expansion of the universe had been discovered.

When Einstein heard about this, he said that inserting the cosmological constant into his equations had been "the biggest blunder of my life". His equations in their original form had predicted a changing universe (either expanding or contracting) – which would have been an amazing theoretical prediction if he had left the equations as they were – but he had made his universe static by inserting the constant. In 1931 Einstein finally removed the cosmological constant from his equations, which became the theoretical and mathematical framework for the expanding universe concept.

The idea that all of the galaxies in the universe are moving away from us, and at speeds that are proportional to their distances from us, is very striking, and can be misleading. You might at first think that we're 'at the centre', but we're not. The galaxies are not themselves moving in this way through space; instead, in Einstein's theory it is space itself that is expanding, and the galaxies are just going along for the ride. All galaxies throughout the universe are moving away from each other. And the further apart any two galaxies are, the faster they are moving away from each other. Observers in each galaxy may naïvely think that they are at the centre of this expansion, as all galaxies appear to be moving away from them, but there is actually no 'centre'.

To understand this, it is helpful to imagine an expanding balloon which has dots all over its surface. We live in a universe with three spatial dimensions. Imagine instead that you live on the two-dimensional surface of the balloon. To you, it is a flat surface. Now imagine how you see the dots as the balloon is blown up: they are all moving away from each other (and from you), with the rate of separation proportional to the separation itself, but none of the dots is 'at the centre'. There is no 'centre' and no 'outside' in this two-dimensional world – the entire universe is expanding.

The expansion of the universe can be extrapolated back to 'the beginning', when the distances between galaxies would have been zero. The universe was once in an extremely compressed state, and originated in what the famous astrophysicist and cosmologist Fred

Hoyle once facetiously called a 'Big Bang' (he was a proponent of the competing 'steady state theory', which was abandoned several decades ago as the evidence for the Big Bang became conclusive). The Big Bang name has stuck, and the Big Bang theory has been the conventional scenario for cosmology for many decades now. According to this theory, if we extrapolate far enough back into the past, the density of matter and the 'curvature' of space become infinite – a so-called 'singularity'. This is the 'beginning' of the Big Bang. However, while we can determine how long ago it occurred, we can say nothing about 'the beginning' itself, or about any 'before'. This provocative and fundamental issue will be discussed in Chap. 5.

The present age of the universe (the time back to the 'beginning') can be computed from the current rate of expansion and the density of the universe. Recent discoveries and modern satellite measurements, described below, add independent new techniques and precision. The best 'cosmological' determination of the present age of the universe is 13.7 billion years. This age agrees well with that determined by astrophysical methods using stars. Combining the observed properties of the oldest clusters of stars ('globular clusters') with our theoretical understanding of stellar evolution gives ages in the range 11–13 billion years. White dwarf stars gradually cool and fade with time; the faintest white dwarfs can therefore give a measure of age. Radioactive dating has also been used to estimate the ages of old stars. All of these astrophysical methods give results that are consistent with the cosmological age of 13.7 billion years.

Is the Big Bang model consistent with Olbers' Paradox? Yes, because the Big Bang universe is both finite in age and expanding. The finite age means that we can only see a finite number of stars, which have existed for less than the age of the universe. And the expansion has caused the brilliant light given off by the Big Bang to be diluted and redshifted from the optical/infrared part of the electromagnetic spectrum to the millimetre band, where it is observed today.

The expansion of the universe has been established beyond doubt. And the implication is that there was a time when the universe was very small and dense. Even if we cannot say anything about the instant of the Big Bang itself, we can certainly say a lot about this hot, dense phase, as the next two sections will show.

Afterglow of the Big Bang

How can we possibly have any idea of what the early universe was like? In its compressed state 13.7 billion years ago, our universe was extremely dense, hot and uniform. But precisely because it was so hot and uniform, the physics involved would have been simple; it was basically a hot 'soup' of fundamental particles and forces. With only a few variables and virtually no complexity, it is relatively easy to compute the properties of that early phase. The physics of the early universe just a small fraction of a second after the Big Bang was already known to us half a century ago; its properties were being explored in the 1940s, when the atomic bomb was being developed.

A fairly obvious test of the reality of the early hot phase of the universe would be the observation of an afterglow. Even, now, 13.7 billion years after the event, there should still be a cool, fading 'relic radiation' left over, which we might be able to detect. That radiation would have cooled as the universe expanded, and would have been increasingly shifted towards the red end of the spectrum. By now it should be only several degrees above absolute zero, and concentrated at microwave (millimetre) wavelengths.

George Gamow, Ralph Alpher and Robert Herman were studying the early universe in the late 1940s. They computed that the temperature of the relic radiation as observed today (called the Cosmic Microwave Background, or CMB) could be as low as 5 degrees Kelvin (5 K, or $-268°$C). A temperature of absolute zero on the Kelvin scale means zero thermal energy – no motions whatsoever – it's as low as you can possibly go, so 5 K is very, very cold. The Kelvin scale is used throughout this book, but you can always subtract 273 to get degrees Celsius ($°$C).

The definitive detection of the CMB was serendipitous. Arno Penzias and Bob Wilson were working at Bell Telephone Laboratories in New Jersey in the mid-1960s, in part to measure the potential background contamination that could affect satellite communications. They worked hard to reduce any radio noise generated by their equipment, and went so far as to delicately remove two pigeons and their droppings from their antenna. But their measurements still showed an excess of 3.5 K, which they could not account for. Meanwhile, in Princeton, just 40 km away, Robert Dicke and his colleagues were using a small radiotelescope to search specifically for the CMB. When Penzias was eventually informed about Dicke's work, he phoned him immediately in puzzlement about his results, and after the phone call Dicke said to his team "Boys, we've been scooped!" Penzias and Wilson were awarded the Nobel Prize in physics for their momentous discovery.

The temperature of the CMB has now been measured with extremely high accuracy. It is 2.725 K. The radiation is constant over the whole sky to an astonishing precision of one part in a hundred thousand. Its spectrum (the distribution of its intensity as a function of wavelength) was found to be very special indeed – it is almost exactly that corresponding to thermodynamic equilibrium, as would be expected for heat radiation coming from an early universe of constant temperature and density (this type of spectrum is called a black-body spectrum, and the observed radiation has the most perfect black-body spectrum known to man). The prediction and discovery of the CMB are considered to be conclusive evidence for the Big Bang theory.

The CMB that we see is an image of the relic radiation as it was at a very specific epoch. This is why it has such well defined properties, rather than being a blur. It comes from the 'surface of last scattering'. Before that time, the radiation (carried by massless particles called photons) scattered off electrons (negatively charged particles of matter), producing an opaque fog. When the universe had expanded and cooled to about 3,000 K, the electrons were able to combine with positively charged protons to form electrically neutral hydrogen atoms, in a process called recombination. This resulted in a decoupling of matter and radiation, and the radiation was finally able to travel freely through space. The universe became transparent. Thus the CMB we see is a snapshot of the surface of last scattering, which occurred 380,000 years after the Big Bang. It's like a distant wallpaper covering the entire sky behind all the stars and galaxies. This is important, as it means that we can clearly see any structures (irregularities in the distribution of matter) that may be imprinted on it. They would show up as a pattern of regions, some very slightly warmer and others very slightly cooler than the average temperature.

And structures there had to be. It was widely believed that the CMB could not be perfectly smooth. If it were, then no galaxies, stars or planets could ever have formed in our universe. But if there were even tiny irregularities in the distribution of matter in the early universe, the slightly denser regions could accrete matter from less dense regions by simple gravitational attraction (even though the overall universe was expanding). They would increase in mass and density and decrease in size until, after hundreds of millions of years, they became so massive and dense that they formed the galaxies, stars and planets that we know today.

What could be the origin of these irregularities – these so-called 'primordial fluctuations'? The most widely held view is that these were random 'quantum' fluctuations (described in Chap. 4) from the very, very early universe that were stretched to macroscopic scales by a brief period in which the universe expanded by an enormous factor. These initial fluctuations grew with time, and became the seeds of an extraordinary pattern that evolved. Overlapping density waves were produced, similar to the overlapping ripples on a pond when a handful of pebbles is thrown in. A pattern caused by overlapping shells is not necessarily easy to see straightaway, but it can be clearly detected using statistical analysis. Cosmologists were able to predict the characteristic size of the waves, which would provide a 'standard ruler' for length scales in cosmology: about 0.6 angular degrees in the CMB, which in today's universe corresponds to about 500 million light-years (5 billion trillion kilometres).

These are amazing predictions. Could they be verified by observations? After many attempts the fluctuations were finally detected in a statistical analysis of an all-sky survey made using NASA's Cosmic Background Explorer satellite in 1992, at a level of 10^{-5} (one hundred thousandth) of the total intensity (another Nobel prize). It was fortunate that the fluctuations were found at this level; if they were much fainter they would have been swamped by fluctuations in the interstellar medium of our own galaxy, and we would never have known about them. As it turned out, they have become a treasure trove of information about the early universe and its large scale properties.

The statistical detection was obviously tantalizing, and many scientists were eager to measure the fluctuations in detail. Several

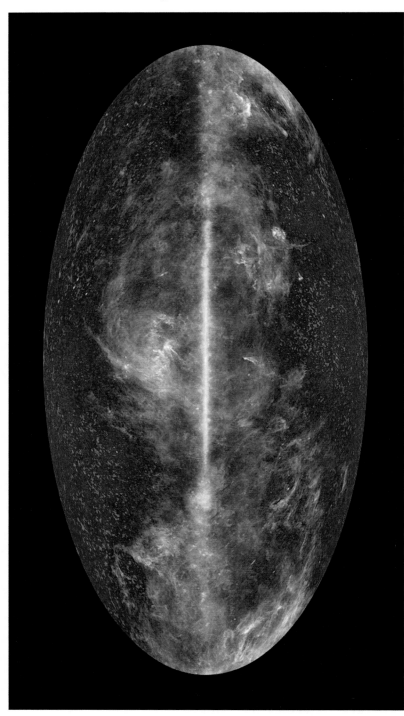

One of the most amazing images ever made of the sky. This is an all-sky image obtained by the Planck team using the Planck spacecraft of the European Space Agency (ESA). The plane of our own galaxy (the Milky Way) is the thin strip extending across the entire image. Above and below it are plumes and filaments belonging to the galaxy. The mottled red background at the top and bottom is the afterglow of the Big Bang, seen as it was when the universe was just 0.003% of its present age. The orange 'spots' are primordial structures that later evolved into stars and galaxies.

hastily built ground based telescopes were able to detect the strongest 'acoustic peak'. A particularly intriguing experiment was a balloon-borne instrument called Boomerang that drifted with the air currents circling the South Pole while staring at one region of sky for a long period of time. It succeeded in making a high-sensitivity map in which the individual fluctuations and patterns were actually visible for the first time. During this time NASA was busy building a follow-up mission: the Wilkinson Microwave Anisotropy Probe (WMAP). Launched in 2001, it has mapped the microwave sky with unprecedented sensitivity and resolution. The detailed agreement of the predicted and observed ripples in the CMB is absolutely astonishing. Even finer detail and new observational horizons will become possible with data from the European Space Agency's Planck spacecraft launched in 2009.

Stimulated by these astounding discoveries, astronomers have searched huge, uniform databases from ground-based surveys of millions of galaxies looking for the equivalent of the CMB ripples in the distribution of galaxies in the 'local' universe. They have succeeded in finding that there is indeed an excess of galaxies separated by the distance (500 million light-years) that corresponds to the cosmological ripples. This provides an amazing link between the local universe and the very distant universe, and a stunning confirmation of predictions made about the early universe and its evolution to the present.

Yet another test can be provided by measurements of the CMB temperature when the universe was at intermediate ages, older than when the CMB was formed (when its temperature was about 3,000 K) but younger than it is today (when its temperature is 2.725 K). The temperature at a given epoch can be measured by observing the ratios of certain atomic and molecular spectral lines. It should decrease at the same rate that the universe expands, and the few measurements made so far seem to indicate that it does.

An added bonus of the observations of the CMB is the accurate measurement of the motion of our galaxy relative to the distant universe. The CMB is a bit warmer on one side of the sky than it is on the other. This is known as the CMB dipole anisotropy. The difference between the two hemispheres is 0.003 K (one thousandth of the total intensity of the CMB) – small, but huge in comparison with the fluctuations described above. It is due to a global Doppler effect – the fact that our local group of galaxies is moving at 627 km per second relative to the reference frame provided by the CMB, so one side of the sky appears redshifted and the other side blueshifted. This motion is a result of the gravitational attraction of other relatively nearby galaxies. Motions such as these in the local universe are being mapped in three dimensions using large galaxy surveys.

Creation of the Elements

The elements, essential for life as we know it, were created in two totally different epochs: (1) in the first minutes, across the entire universe, and (2) billions of years later, in the cores of stars.

When the very early universe was undergoing rapid expansion and cooling, it went through a fleeting moment when the conditions were similar to those in the interiors of stars, and elements could form. That fleeting moment started just seconds after the Big Bang, and lasted just minutes. The process is called Big Bang nucleosynthesis (BBN). Nucleosynthesis is the process of synthesizing the nucleus of one atom from the nuclei of others. Only the first few elements in the periodic table (the so-called 'light elements' or 'primordial elements' – hydrogen, deuterium, helium and lithium) could be formed in this period.

BBN gives the only explanation for the abundances of the light elements. Stars can produce only about one tenth of the helium present in the universe today. Deuterium is actually destroyed in stars. And the BBN predictions are very clear and precise. Again, it was Gamow, Alpher and Herman who first studied this in detail.

The physics of this early phase is well understood. It started when the universe was about one second old, when its temperature was down to 10 billion degrees, at which point stable atomic nuclei could form from the binding together of their constituent particles, protons and neutrons (a neutron is a subatomic particle with no electric charge and a mass slightly greater than that of a proton). Hydrogen nuclei are just protons; deuterium nuclei formed from the fusion of protons and neutrons, and helium nuclei from the fusion of deuterium nuclei.

This phase of the universe lasted for just minutes, and then it was gone. The window of opportunity for the production of elements in the early universe had then passed. The universe continued to cool as it expanded, and the temperature was no longer sufficient to support nucleosynthesis. No elements heavier than helium were able to form, except for trace amounts of lithium, so the light elements were the only ones in the universe for a very long time. The heavier elements would have to wait until the first stars formed, hundreds of millions of years later.

Overall, the abundances (by mass, relative to hydrogen) predicted for the primordial elements were 25% helium, 0.01% deuterium, and 10^{-10} (a tenth of a billionth) lithium. The relative abundance of helium is determined purely by the physics, and is independent of the initial conditions of the universe. It is an extremely robust prediction, and is just what we see in the universe today. The predicted abundances of the other primordial elements also agree well with observations of the universe today.

A variety of methods has been used to measure these abundances. Helium lines are easily seen in the spectra of stars, emission nebulae and planetary nebulae. Deuterium can best be studied by observing isolated gas clouds in the distant universe that are themselves almost primordial, and do not contain stars. These are intergalactic clouds, and we can study them by observing the absorption they cause in the spectra of even more distant quasars that happen to lie behind them. Lithium can be studied in the spectra of old stars, but this is somewhat less certain because of processes in the stars themselves.

There is an important overall check. The primordial element abundances should all be consistent both with each other and with a key cosmological parameter that is related to the density and temperature of the early universe. This parameter has been determined to high accuracy using the WMAP observations discussed above, and it agrees well with the predicted and observed abundances of the primordial elements.

Finally, a recent and important confirmation that the helium was formed in the very early universe, long before the first stars existed, also comes from the WMAP observations. The effect of the primordial helium shows up in the fluctuations of the CMB. As the CMB is observed as it was 380,000 years after the Big Bang, and the first stars weren't formed until the universe was hundreds of millions of years old, this provides direct supporting evidence that the helium was indeed formed in the early universe.

It is amazing to think that we can make such precise and verified statements about such an early phase in the history of the universe – just minutes after the Big Bang. Keep in mind that, while we look back in time as we look out into the universe, we can only see as far back as the microwave background. The universe at ages less than 380,000 years is totally opaque to us across the entire electromagnetic spectrum – we can't see it at all. The short phase when the primordial elements were formed, just minutes after the Big Bang itself, is completely shrouded from our view. Nevertheless we can determine the events of that time from our knowledge of physics, and we know exactly what happened, as proven by abundances measured in the universe today. An incredible success.

However, today's universe contains more than just the light elements. The rest of the elements – the 'heavy elements' – are made in stars.

In total, 94 naturally-occurring elements exist on Earth. Some of these are (in order of the atomic number, which is the number of protons in the nucleus of an atom): hydrogen, helium, lithium, beryllium, boron, carbon, nitrogen, oxygen, fluorine, neon, sodium, magnesium, aluminium, silicon, phosphorus, sulphur, ... iron, ... gold, ... uranium, ... plutonium, ...). Iron is special in that it has the lowest mass per nuclear particle of all the elements. This means that the fusion (joining together) of the nuclei of light elements to make heavier ones (which produces energy) can only work up to iron, because beyond that point no further energy is released. Similarly the fission (breaking apart) of the nuclei of heavy elements to make lighter ones (which also produces energy) can only work as far down as iron. This is a fundamental distinction in nucleosynthesis. Hydrogen bombs are based on the fusion process, and atomic bombs are based on the fission process.

How do stars create elements? Fred Hoyle first outlined the overall process of nucleosynthesis in stars in 1946. A star is born from its parent molecular cloud when its central temperature and density are sufficient for hydrogen fusion reactions to begin. The contraction of the protostar then comes to an end, and a long-term balance is achieved between the force of gravity pulling inwards and the pressure from the hot star pushing outwards. This balance produces the virtually constant, sharply-defined bright sphere that is a star. The star has then reached maturity and has become a so-called main-sequence star.

What determines the minimum and maximum masses of stars in the first place? The lower end is determined by the requirement that the core temperature reaches the 10 million Kelvin necessary for nuclear fusion. If a protostar has a mass less than 0.08 solar masses, it is prevented by basic physics from collapsing sufficiently to reach the required temperature. The result is a brown dwarf star. As 0.08 solar masses is only about 80 times the mass of Jupiter, these objects fill the 'gap' between planets and stars. The upper end of the stellar mass scale is caused by outward radiation pressure in huge and luminous stars ultimately overcoming the gravitational infall of matter. The most massive stars are well over a hundred solar masses. Such massive stars are rare, representing less than one in every hundred thousand stars.

As the lifetimes of stars and the various changes and events that occur throughout their lives depend crucially on their masses, to make things simple we consider just the two extremes, low-mass stars and high-mass stars, with initial masses of less than two solar masses and greater than eight solar masses respectively. A star like the Sun has a main-sequence lifetime of about 10 billion years, compared with just millions of years for very massive stars (perversely the stars with the most fuel have the shortest lifetimes, because they burn fastest).

We start with the low-mass stars. Stars on the main sequence live relatively steady and uneventful lives. They spend their time converting hydrogen to helium. This involves bringing positively charged protons together into the same nucleus, which is not easy because similarly-charged particles repel each other. A very high temperature in the star's core is required to make this possible. In that environment particles are moving in high-speed chaos, and sometimes come close to each other in spite of the electromagnetic repulsion. When they are close enough, another force, called the strong force, overwhelms the electromagnetic force and binds the two particles together. Aside from converting hydrogen to helium, this process results in the net production of energy, because a helium nucleus is slightly (0.7%) less massive than the original four hydrogen nuclei that made it, and the mass difference is converted into energy in accordance with Einstein's famous equation $E = mc^2$, which states that mass is equivalent to energy.

A self-sustaining balance prevails between gravity, the energy produced in the core, and the energy released into space. Eventually, however, the hydrogen in the stellar core becomes depleted, the core begins to shrink, and the star begins to move off the main sequence. The core is now almost entirely helium, but its shrinkage permits the surrounding shell of hydrogen (which is also shrinking) to become hot and dense enough to start hydrogen shell burning ('burning' in this context always means nuclear burning – conversion into other elements through either fusion or fission; in this case it is fusion from hydrogen to helium). This proceeds faster than the main sequence hydrogen burning, and causes an increase in thermal pressure which expands the outer layers; the star becomes a subgiant, and eventually (after a billion years) a red giant. The outer layers of the star experience a weaker pull of gravity, and large amounts of mass escape in stellar winds. This situation persists until the still-shrinking core reaches a temperature of 100 million Kelvin, hot enough for helium-to-carbon burning.

Helium burning starts another and dramatic phase in the star's life. It heats the core excessively, releasing a huge amount of energy in what is called the helium flash. Within seconds the situation is 'corrected', and the total energy production falls sharply. The star becomes a 'normal' helium-burning star, and begins a quieter phase as a horizontal branch star. When the core has been totally transformed into carbon, shell burning around the core (this time of both helium and hydrogen) again causes the star to expand. The helium burning causes a number of thermal pulses of the star during a new red giant phase, and more mass is ejected from its outer envelopes. However, these stars are coming to the end of their lives, as they can never reach the temperatures of more than 600 million Kelvin required for fusion reactions in their carbon cores.

The large sizes of these stars mean that gravity has only a weak hold on the outer layers, and large amounts of matter flow out with the stellar wind. Strong convection during the pulses from the carbon cores dredge up large amounts of carbon, creating what are called carbon stars. The winds from these stars are by far the most prolific producers of carbon – essential for life as we know it.

The matter ejected by these winds, forming what are known as planetary nebulae, disperses into the interstellar medium within a million years. What is left of this star's eventful history is nothing but the cooling carbon core of the star. This remnant is a white dwarf star. It will eventually cool over the far distant future, and ultimately just disappear from view.

High-mass stars are even more spectacular than their low-mass counterparts. Their lives may be short, but they are certainly exciting. High-mass stars are also extremely important, as only they can produce the full range of heavy elements that our lives depend on. They begin their lives just as low-mass stars do, but the nuclear processes are somewhat different and much faster, because of the higher temperature and pressure. The hydrogen in the core is consumed in just several million years, and the subsequent helium burning lasts only a few hundred thousand years. Successive elements are burned more and more quickly, with the core shrinking, surrounded by shells of different elements, and the outer layers continuing to inflate to supergiant scales. The reactions can become quite complex, with heavy nuclei fusing with each other, leap-frogging the buildup to heavier elements. Neutrons can be released, fusing with heavier nuclei to form some of the rarest and heaviest elements. The final result of this frenetic process is the buildup of iron in the core.

As mentioned above, iron is unique and critical, as neither its fusion into heavier elements nor fission into lighter elements releases energy. An iron core means no more energy output. Catastrophic core collapse takes place, followed immediately by a supernova explosion that enriches the surrounding interstellar medium with all the newly created elements. While iron is the heaviest element that can be formed in the usual processes of stellar nucleosynthesis, the extreme conditions in this brief but violent explosive nucleosynthesis create most of the elements heavier than iron. The temperatures reached in these explosions are higher than those in any star, and processes such as neutron capture (which require energy input) create the very heavy elements. Without supernova explosions we wouldn't have most of the copper, silver, iodine, platinum, gold, lead, uranium, and many other heavy elements that are part of our daily lives.

Stellar nucleosynthesis is established beyond doubt. The processes all follow directly from well-known nuclear physics, and many if not most have been verified directly or indirectly from man-made thermonuclear explosions. Detailed and sophisticated computer modelling gives excellent agreement with the wide range of types exhibited by the stars themselves. Observational evidence includes the detailed distribution of the elements across the Periodic Table, including the excess of nuclei with even numbers of protons as predicted, and the fact that young stars contain more heavy elements than do old stars (the oldest contain very little). The heavy elements comprise about 2% of the total mass in the universe today, up from zero percent before the first stars existed: the buildup of heavy elements by nucleosynthesis in stars is clear.

Thus, as we have seen, while the light elements were produced in the Big Bang, the heavier elements are produced in stars. The heavier elements accumulate in the interstellar medium, so stars born later start out already containing elements from previous generations of stars. In this way the heavy elements are continually built up over time. These elements include those of organic chemistry (carbon, nitrogen, oxygen, etc.). The very atoms we're made of were created either in the early universe or in stars.

Evolution of Galaxies from the Dark Ages to Here and Now

We can now study galaxies at all stages of evolution directly. Because of the finite speed of light, we can see the entire history of galaxies laid out on the sky in front of us in one glance. We are presently simultaneously seeing galaxies as they were 13 billion years ago, 10 billion years, 6 billion years ago, 5 billion years ago, 3,2 and 1 billion years ago, 300,000 years ago, 50,000 years ago – all of these at once. Absolutely every epoch, simultaneously, right now in the sky, right there in front of us.

With huge samples of millions of galaxies, astronomers are now mapping out the detailed history of galaxies – every billion years, every million years, whatever interval you choose – the whole history of the universe spread out on a single graph. It's fantastic. 'Detailed history' here means the evolution of the stars in galaxies, the evolution of the gas in galaxies, the evolution of the structure of galaxies, how the galaxies interacted with each other as a function of cosmological time, and many other properties.

By studying similar types of galaxies at different epochs and different types of galaxies at the same epoch, we have multiple views of the history of the universe. Putting it all together is an absolute feast – and it has just become possible over the last two decades thanks to deep surveys of the sky that reach some of the earliest galaxies, novel and clever types of surveys of galaxies at intermediate epochs, and huge surveys at relatively recent epochs covering large areas of the sky. The HST, large ground-based telescopes, super-powerful instruments on many telescopes that can observe thousands of galaxies simultaneously, and super-powerful computers have made this possible. It is truly another scientific revolution.

From this chapter it should be abundantly clear that the universe is indeed expanding and evolving. The observed expansion, the abundances of the light elements, the relic radiation with its near-perfect blackbody spectrum, the fluctuations in the microwave background that led to the formation of stars and galaxies, the production of the heavy elements in stars, and the detailed evolution of galaxies that is now observed across the history of the universe – all of these provide incontrovertible evidence for the expansion and evolution of the universe. The first three have long been referred to as the pillars of the Big Bang Theory, and certainly more pillars now come from our knowledge of the evolution of stars and galaxies. All this gives us great confidence in our understanding of the physics of the universe at an early period in its history, and in Big Bang cosmology in general.

3. What's the Matter?

But in spite of the outstanding successes outlined above, we now find that we do not know what 95% of the universe is made of.

The average density of the universe determines many of the overall properties of the universe. The density to which all measurements are compared is the so-called critical density. It is a tiny 10^{-29} grams per cubic centimetre (about five hydrogen atoms per cubic metre), but it is a central number in cosmology. It is the dividing line between the three main possible geometries of the large-scale universe. If the actual density is less than this, the universe is called 'open': it is negatively curved like a saddle, and parallel light rays diverge from each other over large distances (note that this is a two-dimensional analogue of a three dimensional space, but you get the idea). If the actual density is greater than the critical density, the universe is 'closed': it is positively curved like the surface of a sphere, and parallel light rays converge. But if the actual density is exactly equal to the critical density, the universe is 'flat': parallel light rays remain parallel, as in our ordinary conception of space.

Dynamically as well, everything is compared to the critical density. More matter means more gravity, and that will slow down the expansion of the universe. If the density is less than the critical density, the universe will expand forever, and if the density is greater, the universe will ultimately re-collapse. In a critical universe the amount of matter is exactly that needed to balance the expansion. The rate of expansion just slows with time, and never completely stops. If the universe is on the critical line, it will stay so forever. (All of this assumes that the universe contains only matter – no complicating additional factors such as Einstein's cosmological constant – but stay tuned). There are strong arguments, both theoretical and observational (outlined below), that the total density of the universe is very close to the critical value, and that the universe may well have a flat geometry.

P. Shaver, *Cosmic Heritage*, DOI 10.1007/978-3-642-20261-2_3,
© Springer-Verlag Berlin Heidelberg 2011

How do we actually measure the total density of matter in the universe? The most obvious way is to count all the galaxies in the universe, and multiply by the mass in each galaxy. Galaxies are made of stars. The more stars they have, the brighter they appear. As we know the masses of stars, we can determine the masses of galaxies, and ultimately the total mass in the universe. In addition, we can see that there is luminous matter between the stars (such as the nebulae mentioned above), and we include this too. Even then however, the total density computed for all the luminous matter in the universe comes out to be less than 1% of the critical density.

We also know that some of the matter in the universe is nonluminous (for example, diffuse gas distributed between the galaxies, ghostly particles called neutrinos, and black holes). We can estimate the density contributed by these in various ways, using observations at many wavelengths, from radio to X-ray. This nonluminous matter has been estimated to contribute about 4% to the total density of the universe. So the grand total of the known luminous and nonluminous matter (referred to here as 'ordinary' matter) still only accounts for about 4–5% of the critical density.

There is a totally different way of determining the density of all of this matter. As we saw in the last chapter, primordial nucleosynthesis occurred in the first minutes of the universe and produced the light elements. The speed of this process depended on the density of ordinary matter at that time. The fact that some deuterium nuclei still exist in the universe today tells us that this process stopped before all the deuterium nuclei were used up, and the measured abundance of deuterium in turn tells us the density. The result is consistent with that determined above. Thus, two independent methods both indicate that the total density of ordinary matter is well below the critical density. Are we missing something?

Dark Matter

In the 1930s, Fritz Zwicky was examining the motions of the outermost members of the nearby Coma cluster of galaxies and noticed that they were moving far more quickly than could be accounted for by the gravitational attraction of the luminous matter in the cluster. With their rapid motions they could not be physically bound to the cluster – they would be flung out away from it and dispersed. But the cluster, including these outlying galaxies, was clearly a physical entity. To explain this paradox, Zwicky hypothesized that there may be a large amount of unknown 'dark matter' in the cluster that made up for the missing mass. This dark matter would actually have to contain most of the mass of the cluster.

By the 1970s the evidence for dark matter was much stronger, particularly due to the work of Vera Rubin and colleagues who showed that the same argument applied to individual galaxies. The outermost stars could only be kept at their high orbital velocities around the spinning galaxies if the visible galaxies were surrounded by giant spheres of unseen dark matter, which accounted for the bulk of the mass of each galaxy. It is now known that the total mass of dark matter in our universe is over five times the mass of the ordinary matter we know about (both luminous and nonluminous). Our conventional view of the universe is grossly distorted: the ordinary matter we actually see is just froth on a giant ocean of unknown dark matter.

We have no idea what this dominant form of dark matter is, although we have some idea of its overall properties. In order to be bound in and around galaxies, it cannot be comprised of particles that move too quickly – otherwise it would be spread out and free stream more evenly throughout space. Therefore one talks about 'cold dark matter'. Could it be ordinary matter which is simply not luminous and had not been counted in the above census, such as neutron stars, brown dwarfs or free-floating planets, or even more black holes? Such possibilities have collectively been given the entertaining name MACHOs (Massive Astrophysical Compact Halo Objects). MACHOs would individually be hard to detect, but determining whether large numbers of them comprise the dark matter is easier. The trick is to search for cases in which the light of a background star is temporarily brightened by gravitational lensing when a MACHO passes directly in front of it. This has been done by several groups by observing millions of stars every night and looking for light variations using powerful computers. The results rule out the possibility that MACHOs comprise a significant fraction of the dark matter.

The nucleosynthesis limit, which refers to all ordinary matter, also excludes any significant amount that we had initially missed. That means that, if there is indeed a large amount of 'missing mass', as indicated by the work of Zwicky, Rubin and others, it cannot be ordinary matter. (A totally different possibility is that there is no 'missing mass' problem at all; the observed effects might be due to the law of gravity on large scales being different from the standard Newtonian or Einsteinian laws that we know so well on local scales, rather than any missing mass. But most physicists don't want to tinker with the laws of physics, and in any case this proposal seems unlikely given what we know about the large scale properties of the universe today.)

The 'missing mass' must be fundamentally different from the ordinary matter of the universe. Physicists have rushed in with a host of possible candidates as explanations for this 'exotic' form of dark matter. The favourites are hypothetical particles called WIMPs (Weakly Interacting Massive Particles). Finding such particles is an extremely high priority in both physics and cosmology, and searches of all kinds are going on. They may be everywhere – in space, in the Sun, in our atmosphere, streaming right through the Earth and right through us – so searches are going on everywhere: the outskirts of galaxies, in vast underground experiments and in particle accelerators. The search for exotic dark matter has provided a huge stimulus to experimental physics, and may turn up totally unexpected new particles and phenomena in addition to the dark matter itself.

However, even including both the ordinary and exotic components, the total density of matter in the universe is still only about a quarter of the critical density.

Dark Energy

Even more surprising than exotic dark matter was the discovery of dark energy. In the late 1990s, two large independent groups of researchers were trying to study the evolution of the universe over a large fraction of its history by comparing supernovae occurring in nearby and distant galaxies. The parameter they were trying to measure was the long-sought *deceleration* parameter. As it turns out they discovered *acceleration* instead!

The technique involved measuring two numbers at each of two different epochs: the rate of expansion of the universe, and the distance from us. The rate of expansion is the easy part: it is given by the redshift measured at each epoch. Measuring the distance is much more challenging, but simple in concept. Imagine comparing the light you see from a 60 W lightbulb held at arm's length with the light you see from the same lightbulb when it is 1 km away. At a greater distance it appears much fainter, and if you know that it is the same lightbulb you can easily calculate its distance (on our familiar local scales, if it is four times fainter it is twice as distant). Something with such a constant intrinsic luminosity is called a 'standard candle'. But can we find such a wonderful thing in the real universe?

Amazingly, 'messy' supernova explosions seem to provide the answer. Supernovae of type Ia all have essentially the same intrinsic luminosity, and they can be seen across the entire universe. Perfect. But how can they all have the same luminosity? According to fundamental physics white dwarf stars cannot be more massive than 1.44 solar masses; any star more massive than that will collapse due to gravity. If a white dwarf star is in a binary system with a younger star whose outer envelope escapes and descends onto the white dwarf, the white dwarf star has no choice but to explode as a supernova as its mass approaches the critical limit. The magnitude of the explosion is absolutely determined by the limit of 1.44 solar masses: it is a type Ia supernova, virtually identical to all other type Ia supernovae anywhere in the universe. This is determined by fundamental physics, and is unavoidable. Therefore, type Ia supernovae may be the ultimate standard candles.

Do observations support this? The spectra of these supernovae, near and far, seem to be identical. Their 'light curves' (the rate and shape of their declines) are identical. Their detailed spectra (and therefore chemistry and dynamics) are identical. Any normal intervening dust that might diminish the apparent brightness of the distant supernovae would show up by causing their overall spectra to be redder (the red end would be diminished less than the blue end). Type Ia supernovae do indeed seem to be excellent standard candles, and, if so, they are ideal for studying the large-scale dynamics of the universe.

From observations of large numbers of these supernovae, both near and far, the two groups of researchers independently discovered that, relative to nearby supernovae, the most distant supernovae are significantly fainter than they would be in a decelerating universe, and concluded that the expansion of the universe over the last several billion years has actually been accelerating.

The cause of this acceleration is unknown. In our ignorance it has been attributed to a mysterious 'dark energy', which has a repulsive force across the universe.

Physicists were shocked. In their view any such force could only be immensely huge or exactly zero (see next chapter), but certainly not the small but finite value indicated by the observations.

Cosmologists, on the other hand, were delighted. Recall that, from Einstein's theory of relativity, mass and energy are equivalent. The newly discovered dark energy brings the total mass-energy of the universe very close to the critical density corresponding to a flat universe, which is one of the major predictions of the popular inflationary cosmology discussed in Chap. 5.

Precision Cosmology

Notwithstanding the surprises arising from the discoveries of exotic dark matter and dark energy, observationally the large-scale properties of the universe are now considered to be rather well known, and it is commonly said that we have entered a period of 'precision cosmology', in the sense that we know the parameters that describe our universe with considerable accuracy, even if we do not yet know some of the major underlying causes. Perhaps exotic dark matter is inert aside from its gravitational effects, and dark energy may be just a single constant. The number of parameters that describe today's cosmological model ranges from 4 to 20, depending on the analysis and type of data set used.

The now so-called 'standard model of cosmology' is the favourite. Here are some of the parameters, to give a feeling for the accuracy achieved. The age of the universe is 13.7 billion years.

The curvature of space is within 1% of a 'flat' geometry for the universe. Just 4.6% of the total mass-energy of the universe comes from normal atomic matter that comprises galaxies, stars planets and us, 23.3% comes from exotic dark matter, and 72.1% comes from dark energy. Recombination (the era when the microwave background photons were emitted) occurred when the universe was 380,000 years old. Other parameters put strong limits on the nature of the dark energy, describe what the primordial irregularities were like that ultimately formed the galaxies and stars, and tell us when later major epochs in the history of the universe occurred. It has been a bonanza. The most precise data to date have come from WMAP, and even better will come soon from the European Space Agency's new Planck spacecraft.

Remarkable as they are, these results may be considered to be part of 'classical cosmology'. Because of developments in physics over the last century, we must view the early universe in a fundamentally new way. Einstein's relativistic cosmology is one thing, but the extremely small scales of the very early universe demand an entirely new way of thinking, as will be shown in the next chapter.

4. Inner and Outer Space

Studies of the very small and the very large have now merged – in cosmology.

Particle Physics and Cosmology Merge in the Big Bang

For centuries we have been probing ever deeper into the nature of matter. Within the last century we entered the world of atomic physics, then nuclear physics, then elementary particle physics. How small are we talking about here? Atoms are typically about 10^{-11} metres in size, and atomic nuclei about 10^{-15} metres. Elementary particle physics reaches down to 10^{-18} metres and even smaller. That's less than a millionth of a trillionth of a metre.

At the same time we've been exploring ever larger scales in our universe – more than 10^{+24} metres. Only a century ago we thought that the entire universe was nothing more than our own galaxy. Now we know that the universe is vastly bigger, containing over a hundred billion galaxies. Studies of the very small and the very large have been moving in opposite directions.

But we also now know from the Big Bang theory that our entire universe was once very small – even smaller than the scale of elementary particle physics mentioned above. Thus, in the early universe, subatomic physics was intimately connected with the properties of our entire universe. To study the universe we have to know all about physics on the smallest scales. We have to understand fundamental physics. Hence this small diversion from the main story line.

The first thing we should understand about this subatomic world is that it is conceptually very different from our everyday experience. This should not come as a surprise, as our normal

P. Shaver, *Cosmic Heritage*, DOI 10.1007/978-3-642-20261-2_4,
© Springer-Verlag Berlin Heidelberg 2011

experience is limited to our human scale. Even on the atomic scale things are quite different – the hard desk in front of you is mostly empty space, with atoms and molecules tied together by the electromagnetic force. The other fundamental forces of nature are the strong force (which holds atomic nuclei together), the weak force (responsible for radioactive decay), and the gravitational force. As we go to subatomic scales we encounter a wide and exotic variety of particles and antiparticles, and fanciful properties such as 'charm', 'flavour' and 'strange'. The objective has always been to determine which are the truly fundamental particles (those with no substructure – the basic building blocks of all the others).

Good to 1 Part in 10 Billion

The Standard Model of particle physics is the most successful and well-verified theory known to man. According to this theory, the most fundamental constituents of matter are thought to be the families of fermions (leptons and quarks) plus their antiparticles (each particle has an antiparticle with the same mass and opposite electric charge). These include the electron and two similar but heavier particles (the muon and the tau), six different types of quarks (which are the basic constituents of protons and neutrons), and three different types of neutrinos. There are also the bosons, the carriers of the fundamental forces of the Standard Model: the photon carries the electromagnetic force, the gluons carry the strong force, and the W and Z particles carry the weak force. Not be forgotten is the still undiscovered Higgs boson, which is supposed to be responsible for giving particles their masses. (Don't worry about the complicated terminology; the only particles you need to remember are the positively charged proton and the chargeless neutron, which are the constituents of an atomic nucleus, the negatively charged electron, which orbits around the nucleus, the photon, which carries light, and the so-far hypothetical Higgs boson, which is thought to give particles their masses.) The particles are all considered to be point-like, with no spatial extent whatsoever. The Standard Model explains virtually all the results from the world's giant particle accelerators. Some of its predictions have been confirmed to an accuracy of 1 part in 10 billion.

In spite of its enormous success, the Standard Model is still thought to be unsatisfactory because of its many parameters and particles, and the (so far) non-detection of the Higgs boson. It does not explain why these particles exist, and why they have the properties they have. A possible next step would be to unify the electroweak and strong forces into a 'Grand Unified Theory'. An important concept that goes beyond the Standard Model is supersymmetry, in which particles occur in supersymmetric pairs; this would help to unify the fundamental interactions in nature, and evidence of supersymmetry is eagerly being sought. Various other possibilities have been explored, but none has so far been successful.

Search for a Unified 'Theory of Everything'

Ultimately one would like to include the force of gravity, but this is very difficult because unification would involve a fusion of general relativity (gravity) and 'quantum mechanics'. The world of subatomic physics is intimately connected with the concepts and techniques of quantum mechanics. The word 'quantum' refers to a discrete amount of something. Its origin in subatomic physics came with the realization that not only particles, but also waves and other phenomena can occur in discrete amounts. However, Einstein's theory of gravity involves smooth variations in space and time over large distances, the opposite of the abrupt discreteness of quantum mechanics and particle physics.

But are the 'fundamental' (or 'elementary') particles of nature really point-like particles at all? It has been suggested that they may instead be comprised of 'strings' in a ten-dimensional space, with six of the dimensions compactified to immeasurably small scales while the other four dimensions are our familiar dimensions of space and time (which have to be treated together in relativity as a four-dimensional 'spacetime', informally sometimes just called 'space'). The string idea forms the basis of superstring theory (or string theory for short), which is extremely popular today because it may unify all the particles and forces together with the force of gravity into one 'Theory of Everything'. A great many theorists are working to develop this theory, as the stakes are so high: it may provide the ultimate merger of the large and small.

According to string theory there is just one single entity – the string – underlying all of fundamental physics. There is nothing more fundamental. The myriad particles and their properties are just manifestations of the various vibrational patterns that strings can produce. This applies to the force particles as well as to the matter particles. Particle masses are given by the energies of the vibrating strings, and properties such as the electric charge arise from other characteristics of the vibrations. Strings would provide the ultimate unification. Unlike the particles of the Standard Model, the strings are spatially extended; they are tiny one-dimensional filaments of energy that can vibrate in a large number of ways. The fact that they are spatially extended means that there are no 'singularities' or infinities in this theory, which should therefore never break down. And although they are extended along one dimension, the lengths of the proposed strings are so small (perhaps approaching the Planck length, 10^{-35} m, at which the frenzy of quantum fluctuations may become dominant) that they would be the smallest and most fundamental entities possible, the end of the line in the search for ever-smaller constituents of matter.

String theory has had a somewhat tortured history. It originated from efforts made in 1968 to better understand the strong force. In 1970 it was realized that the equations implied a one-dimensional (string) phenomenon, but they did not match experimental results. Meanwhile the Standard Theory of particle physics was enjoying phenomenal success. John Schwarz and others nevertheless persevered, and proposed that the new string theory may give a quantum mechanical description of gravity. In 1984 Schwarz and others made a breakthrough in string theory, and many theorists were immediately attracted to it as the way of the future, pointing to a possible revolution in physics. It was later found that six new dimensions of space had to be incorporated (adding to the original four), but such complications were accommodated, and the advance of string theory continued.

In 1995, Ed Witten found a way of unifying the several string theories that had proliferated by then. M-theory, as he tentatively called it, came at a cost of yet another dimension, but it contains other structures in addition to strings: two-dimension membranes called two-branes, three-dimensional objects called three-branes, and even p-dimensional objects called p-branes. So now we have a master theory with 11 dimensions (10 of space and 1 of time).

For all their promise, string theory and M-theory have their detractors. After several decades of (somewhat haphazard) development, they have so far had no experimental verification. There are several other issues. A good scientific theory should be economical, and some researchers consider the extra dimensions to be extravagant complications to a Theory of Everything. The sizes of the extra dimensions are still uncertain. And it has been estimated that there are some 10^{500} possible ways of folding them – 10^{500} possible solutions to string theory. The strings provide far too many vibrations, almost all of them outside the range of the most powerful particle accelerators. At best, string theory will only be accessible through the lowest-energy (massless) vibrations. The masses of the known particles do not seem to fit the string scenario. A desired characteristic of a fundamental Theory of Everything is that it can be formulated free of the standard concepts of space and time, and that has not yet been achieved. And generally, some researchers think that something so removed from experiment lies outside the domain of science. Nevertheless, while it is still very much a work in progress, the potential payoff of string theory (and M-theory), explaining all of particle physics and merging quantum mechanics and general relativity into a fundamental Theory of Everything, is so great that a huge effort, involving thousands of physicists, is going into it.

String theory and M-theory have a competitor, by the name of loop quantum gravity. This theory, first developed in the mid-1980s, is another possibility for merging general relativity and quantum mechanics. The two theories approach the same objective from opposite points of view: the small and the large. String theory arose from particle physics and later encompassed gravity; loop quantum gravity had general relativity as its starting point, and seeks to incorporate quantum mechanics. It is conceivable that the two approaches will eventually come together in establishing the basis for a Theory of Everything.

It is hoped that the huge new particle accelerator, the Large Hadron Collider (LHC) at CERN (the European Organization for Nuclear Research) near Geneva, will make major discoveries relevant to these theories.

Conditions that existed in the very early universe are reproduced using collisions of lead ions in CERN's Large Hadron Collider. The event shown here was recorded in late 2010 by the ALICE experiment at the LHC.

Quantum Wierdness

Quantum mechanics has been a fundamental part of physics for most of the past century, but some of its concepts are bizarre to say the least.

For example, a sub-atomic entity can exhibit the properties of both a particle and a wave at the same time; this phenomenon is called 'duality'. However, when an observation is made, the entity is found to be either in one state or the other. This has led to some of the most famous debates in the history of physics, and even to suggestions that the entity may be in all possible states at once, co-existing in 'parallel universes'. This was one of the precursors to current ideas about a 'multiverse', discussed below.

Undoubtedly the best known feature of quantum mechanics is Heisenberg's uncertainty principle, which states that if one property of a particle (e.g. its position) can be accurately determined in a given measurement, then a corresponding property (its momentum) cannot. This is not a result of measurement error, it is inherent in the physical world. It is a fundamental and unavoidable feature in quantum mechanics, and its validity has been extremely well confirmed by experiments.

The wave aspect of a physical entity refers to the probability of one of its properties having a given value; it is called a probability wave, or wavefunction. The wavefunction of an entity exists everywhere – it describes the probability at every possible location. If a particle has a well-defined position, the corresponding wavefunction is strongly peaked at that position. Furthermore, all properties of physical entities exist in pairs that are 'duals' of one another. Examples are position and momentum, and time and energy. This duality leads naturally to an uncertainty relation between them: if the position is sharply defined, the momentum cannot be. The uncertainty has the remarkable property that it is intrinsic in nature. That is, it exists quite independently of us and our measurement apparatus.

Consider determining the position and momentum of a particle by illuminating it with light. To determine the position accurately, we cannot use light with long wavelengths, as that would blur the image. We have to use light with the shortest possible wavelengths. But light is quantized; it is composed of discrete photons. And the shorter the wavelength, the greater the energy of the individual photons. The high energy of these photons disturbs the momentum of the particle, making a simultaneous accurate measurement of the momentum impossible, even in principle. This is inherent in the quantum nature of physical reality.

The name of this principle has been much discussed. Proposed names ranged from those that implied limitations to measurements to others that convey the intrinsic nature of the phenomenon: uncertainty, indeterminacy, inexactness, and others that mean blurred, fuzzy and diffuse. The subtlety of the differences between these names gives an idea of the agony over the exact meaning of the concept. Nevertheless, it is a cornerstone of quantum mechanics and has been hugely successful.

Einstein was always upset by the uncertainty principle, and felt that it indicated that quantum mechanics was an incomplete theory – that someday an underlying reality would be found in which strict causality reigned rather than mere probability. He invented 'thought experiments' intended to show the inconsistency and fallacy of quantum mechanics.

The most famous of these was published in a 1935 paper he co-authored with Boris Podolsky and Nathan Rosen, which became known as the EPR paper. The essence of the paper is as follows. For simplicity, consider an initial particle that disintegrates into two particles of equal mass that fly off in opposite directions. The 'velocities' of the two particles, A and B, are equal and opposite, and their positions are similarly related. Suppose you then measure the velocity of particle A. This automatically tells you the velocity of particle B, without interacting with it in any way, and implies that particle B has a definite velocity at any given moment. You could just as well do this for position. The conclusion drawn by EPR is that particles do have definite positions and velocities. (Note that you could still not accurately measure both the position and the velocity of a given particle, as this would violate the uncertainty principle – but even just measuring one of the two demonstrates that they do have definite values, and this is all that EPR wanted to establish.)

EPR had made their point, although it was not proven, and no one expected that a relevant experiment could ever be carried out.

But in 1964 physicist John Bell proposed an ingenious version of the experiment that could conceivably be made. It involved a quantum property called spin. Without going into technical details, suffice it to say that the experiment was eventually carried out, by Alain Aspect and others, and it produced an absolutely stunning result: the EPR viewpoint was refuted, and the predictions of the uncertainty principle were confirmed. Furthermore, even though quantum properties appear at random, they are correlated in the two particles that emerged from the subatomic event. As nothing travels faster than light, this means that the two particles somehow 'know' the state of each other without information being conveyed – they are 'entangled particles' – essentially parts of the same physical entity, a correlated system, no matter how far apart.

Quantum entanglement, as this astonishing phenomenon is called, has been proven beyond any doubt. The experiment has been repeated many times, and over distances of tens of kilometers. There seems to be no limit to the separation; quantum entanglement could conceivably reach instantaneously across the entire universe. It makes such exotic possibilities as teleportation and quantum computing conceivable, at least in principle (although it still doesn't mean that information can be sent faster than light).

The Quantum Activity of Empty Space

A crucial implication of the uncertainty principle for discussions of the very early universe is that quantum fluctuations can arise from empty space. Energy and time, like position and momentum, are subject to the uncertainty principle. The conservation of energy can appear to be violated, if only for fleeting moments. This allows particle-antiparticle pairs of 'virtual particles' to be created, so even 'empty' space is teeming with virtual particle pairs that can come into existence and then disappear again on extremely short time scales. The uncertainty principle also applies to all fields (such as the electromagnetic, strong, weak and gravitational fields); the value of a field at a given position and how quickly it is changing cannot both be known simultaneously.

So fields are also fluctuating randomly in 'empty space' all the time. All of these together comprise the 'vacuum fluctuations' of empty space.

The quantum activity of empty space has actually been verified experimentally. Hendrik Casimir predicted in 1948 that there would be less quantum activity between two metal plates in otherwise empty space than elsewhere, causing the plates to be pushed towards each other. This extraordinary prediction was later confirmed experimentally to an accuracy of better than 5%. Vacuum fluctuations also have an effect on the magnetic properties of electrons, and in this case theory and experiment agree to one part in a billion. Space is certainly not empty – it's full of quantum activity.

However, in spite of great theoretical progress and experimental verification, it's still not all clear sailing in empty space. The type Ia supernova results described in Chap. 3 have confronted quantum physics with the so-called 'cosmological constant problem'. The dark energy implied by the supernova observations is reminiscent of Einstein's cosmological constant, but it is 10^{120} times less than would be expected from the vacuum fluctuations mentioned above. That is 1 followed by 120 zeros – a huge discrepancy. The quantum theory value comes directly from the uncertainty principle; it is very large because fields have so many modes of vibration. As no dark energy had been detected until 1998, it had been thought (hoped) that the value predicted by quantum theory may be driven to exactly zero by some as yet unknown mechanism, such as a symmetry that cancels out. Now that dark energy has been detected, but is neither zero nor the immensely larger value predicted by quantum theories, it poses a major problem for quantum theory. (It has sometimes been said that the quantum value is "the worst prediction of all time"!). In attempts to solve this discrepancy, alternatives to the cosmological constant have been proposed, such as a scalar field called quintessence which can vary in time and space, but at the moment the odds favour the cosmological constant. Interestingly, Steven Weinberg came up with a (possibly controversial) solution to this problem in a different context before the supernova results were known; this is elaborated in Chap. 6.

In spite of this issue, the quantum activity of empty space is a proven reality, and in the next chapter we will see its relevance to the large scale properties of the universe, to the Big Bang, and to the possibility of myriad other universes. There is no question that the entire universe is intimately connected with the subatomic world.

5. The Origin of Our Universe – and Others?

Can we explain the origin of our universe? And – even more provocatively – are there other universes aside from our own?

A Beginning?

If the universe is expanding, it seems just common sense to extrapolate back and expect to find a time when it had zero size. And indeed, according to the classical Big Bang theory, which is based on Einstein's General Theory of Relativity, as we follow the universe back in its history we come to an instant (now computed to be 13.7 billion years ago) when the dimensions of space were compressed into a single point and time began – a so-called mathematical 'singularity' when quantities such as density were infinite. But according to the theory there was no 'outside' – it was the entire universe that expanded, and it did not expand 'into' anything (consider again the balloon analogy given in Chap. 2). There was also no 'before' – time had its origin in the Big Bang. Period. These are mind-boggling concepts for anyone. How can they be explained?

Einstein's theory has been stunningly successful, and it has provided us with a highly reliable framework for studying the universe from early times and on the largest scales. In this theory, space and time are part of the universe, not merely the theatre within which the universe 'happens'. Space is finite but unbounded, while time is finite and bounded. Space and time are part of one spacetime, so the limiting singularity at the instant of the Big Bang applies to both. Time itself began at the moment of the Big Bang. The theory tells us what happened after the Big Bang, but it says nothing about the Big Bang itself. When any mathematical theory encounters the infinity of a singularity – infinite density, pressure and temperature – its equations break down and don't work at that extreme. So it is with general relativity. While it is commonly said that the singularity was the beginning of time, it can also be said that an extrapolation to infinity is highly dubious. There are no other profound insights which inform us that there is no 'outside' and no 'before'. These concepts are simply implicit in Einstein's general theory, and that theory has worked extremely well.

Most cosmologists have just taken the Big Bang as an unexplained 'given', and concentrated on working out the events in later stages of the universe (and they have been spectacularly successful in doing so – correctly predicting the abundances of the light elements formed just minutes after the Big Bang, and correctly predicting the relic radiation from the Big Bang and the structures in it).

There have been some attempts to make theories that eliminate the beginning of time (remove the singularity of the Big Bang) such as the 'no-boundary proposal' of James Hartle and Stephen Hawking in the early 1980s. They attempted to make a quantum theory of gravity in which the entire spacetime of the universe is finite in extent but contains no boundaries. Without boundaries there would be no need to specify initial (or boundary) conditions; the universe would be completely self-contained. On the superfine scales where quantum effects are important, not only would the individual points in spacetime become blurred, but so too would the very distinction between the space and time dimensions. The time dimension could become spacelike in extreme conditions, and the singularity of the Big Bang would be smeared out, eliminating any well-defined instant of creation. As Hawking once commented, asking what came before the Big Bang is then rather like asking what is north of the north pole.

The message is clear: we don't know how the universe began. We only know how it evolved. But over the years some fundamental questions concerning the properties of the early universe have led to a major revolution in our picture of the Big Bang scenario, and even to new possibilities for the ultimate origin of our universe.

Cosmic Inflation

In the 1970s a variety of serious problems were becoming apparent in the standard Big Bang model. One was the so-called 'horizon problem'. In the standard Big Bang scenario even the speed of light was inadequate to causally connect different regions over the sky. Independent regions could have had very different temperatures, as they had never been in contact with each other, but satellite observations showed that the microwave background has the same temperature over the whole sky to a precision of one part in a hundred thousand. How could this possibly be?

You might think that the horizon problem would have been less severe in the early universe, when distances were smaller. But in fact it would have been worse. Looking back, in order to halve the separation between two regions in the universe we would have to go more than halfway back in time, because gravity slows the expansion of the universe. Light would therefore have travelled less than half as far, and would have been less able to keep the two regions in contact than today. So the horizon problem was even worse in the early universe than it is now.

Another problem (the 'flatness problem') was that the density of the universe today is very close to the critical value (10^{-29} grams per cubic centimetre), the dividing line between an open or closed universe. Extrapolating back in time, this 'coincidence' becomes tighter and tighter. If the density were exactly equal to the critical value, then it would remain equal to the critical value forever. But if it deviated even slightly from the critical value, then it would deviate more and more as time went on. Expressed another way, to be as close as it is to the critical value today, it would have to have been within one part in 10^{60} of the critical density just after the Big Bang.

A third problem of the standard Big Bang scenario was the prediction of exotic particles such as magnetic monopoles, while none have been observed.

To address these problems, Alan Guth introduced the concept of 'inflation' in 1979. He suggested that, at a very early stage, about 10^{-36} seconds after the Big Bang, the universe may have undergone an extremely fast and phenomenal expansion due to a 'phase transition' in response to the rapidly decreasing temperature (a familiar phase transition is that from steam to water, or water to ice). The dimensions of the universe may have increased by more than an incredible factor of at least 10^{30} in just 10^{-33} seconds. That's an increase of over a million trillion trillion in less than a billionth of a trillionth of a trillionth of a second.

This would mean that the patch of the universe that we now observe is only a miniscule fraction of the total. Imagine blowing a balloon up by a large amount. On a given scale its surface would become much, much smoother and flatter. Any irregularities in 'our patch' of the universe would become very small, explaining the extremely constant temperature we see over the sky. And the surface of our patch would become arbitrarily flat (depending on how much we blew up the balloon), explaining how close the density of the universe is to the critical value. Inflation would also eliminate the problem of magnetic monopoles by enormously increasing the volume they could occupy; the probability of even one of them being in our small part of the universe becomes negligible. Thus, inflation would solve some of the major problems of cosmology in one swoop. The inflationary scenario has become widely accepted, and has been honoured with the name inflationary cosmology.

But how did it start and how does it work? Guth didn't profess to know the original trigger, but he commented that once it got going it became the "ultimate free lunch." Something for (almost) nothing? Could this be the secret for the early expansion of the entire universe?

It does turn out to be something pretty exotic: antigravity. The story actually goes back to Einstein (yet again), and his attempt to use his new theory of general relativity to describe the universe in 1917. You may recall from Chap. 2 that he found that his equations implied that the universe had to be either expanding or contracting. There was no solution for a static universe, which was the dogma of the time. But then he realized that his equations could accommodate a constant term, the cosmological constant mentioned earlier, which produces a repulsive force (antigravity) that can counteract the normal attractive force of gravity. The theory did not constrain the value of this constant, so Einstein was free to choose a value that would exactly cancel the gravitational self-attraction of all the matter in the universe. When it was later discovered that the universe is in fact expanding, Einstein removed the cosmological constant from his equations, and it was largely forgotten for decades.

In 1979 Guth was approaching the problems outlined above from a particle physics point of view. Various fields (such as the electric field) permeate otherwise empty space, although an observer immersed in them would not perceive them – space looks and feels completely empty. Some fields, such as the so-called scalar fields, can produce a repulsive force, just like Einstein's cosmological constant. The energy density in such fields is constant, so the total energy is proportional to the volume occupied. This means that if the volume containing such a field were increasing, the energy and repulsive force of the field would also increase. It would quickly overcome the normal attractive gravity of matter, causing a huge exponential expansion. As Alex Vilenkin once commented, Einstein just wanted to balance the universe, whereas Guth wanted to blow it up! The increasing positive energy of the scalar field is balanced by the negative energy of the deepening gravitational potential, so the net energy cost of inflating the universe can effectively be zero (everything has energy according to Einstein's famous $E = mc^2$, but everything also has negative energy due to gravity). This extraordinary period of phenomenal growth comes to an end when, in particle physics parlance, the high energy, unstable 'false vacuum' decays into the low energy, stable 'true vacuum' of everyday life. The energy is suddenly released in a gigantic fireball of radiation and elementary particles, and this is followed by the 'normal' evolution of the universe as described by the standard Big Bang theory (see Chap. 7).

So a huge universe could be quickly produced from a tiny seed. But what was the seed? Perhaps it arose from a phase change in the very early universe, such as that which produced the strong force. Or perhaps, even more speculatively, it might have been a random fluctuation in a pre-existing 'quantum ocean'. We just don't know. Inflation has expanded our theoretical horizons almost as much as it has the universe.

While many aspects of inflationary theory remain speculative, many others have been worked out, and there is no question that the concept of inflation has been a huge success insofar as it has solved many of the outstanding problems of the original Big Bang theory. One issue that at first seemed to be a problem for inflation has turned out to be one of its successes. At the end of the inflationary period, as the energy was released, quantum effects would inevitably have produced irregularities from one position in space to another. Would these fluctuations end up being more of a problem than those that inflation was intended to solve? At a workshop in Cambridge in 1982, four teams independently made the computations, and the results were unanimous. The magnitude of the fluctuations should be nearly the same on all length scales, from that of our galaxy to that of the entire universe. This was a clear prediction of inflation, and it could be tested. Observations made with the WMAP satellite over the past decade have confirmed this prediction.

So Einstein was perhaps a bit hasty in withdrawing his cosmological constant. It, or something very much like it, appears to have been involved in both the current acceleration of the expansion of the universe and the early inflationary expansion of the universe – and perhaps more?

The Bang and 'Before'

Inflationary cosmology has been extremely influential in stimulating radical new ideas. Along with the spacetime concepts mentioned in the last chapter, it has encouraged intrepid physicists and cosmologists to go beyond the normally accepted bounds of the standard Big Bang model. It has inspired speculation about the possible cause of the Big Bang and what, if anything, may have gone before.

When Guth introduced the concept of inflation, it was seen as *an* event in the very early history of the hot Big Bang universe – but not *the* creation event. Over the last few decades that distinction has become blurred. For one thing, the phenomenal events of inflation may have completely obliterated any evidence of what went before. Furthermore, the quantum uncertainty dominating the earliest fraction of a second (i.e. before the Planck time, 10^{-43} seconds after the Big Bang) would surely have created an impenetrable fog. Any knowledge of the ultimate origin of our universe may have been made forever inaccessible.

However, inflationary cosmology has stimulated remarkable new speculations. It, along with quantum mechanics, has led to the suggestion that cosmic inflation may be happening everywhere and all the time – outside our universe! Our entire universe may be just one speck in a huge sea of frenzied quantum foam of false vacuum, with the seeds for endless universes continually being created randomly by quantum fluctuations. As Andrei Linde put it in 1994, "Although this scenario makes the existence of the initial Big Bang almost irrelevant for all practical purposes, one can consider the moment of formation of each inflationary bubble as a new 'Big Bang'. From this perspective, inflation is not a part of the Big Bang theory, as we thought 15 years ago. On the contrary, the Big Bang is a part of the inflationary model". Welcome to the Multiverse!

Here be Monsters

Hundreds of years ago, when cartographers were drawing maps of what was then known about the world, they wrote "Here be Monsters" over the vast uncharted lands and oceans. They sometimes embellished these regions with pictures of dragons, sea monsters, violent storms and sinking ships. Now, 'looking out' through the portals of the inflationary early universe, we can perhaps imagine something similar.

There could in principle be an infinite number of universes constantly being formed at different times – inflating regions each of which ultimately becomes a separate universe. This multiverse, as it is called, could be infinite and eternal, containing endless cycles of different universes, with no beginning in time. Our universe may be just one of a vast number of universes – a 'local patch' in the multiverse.

The multiverse concept has in fact arisen from various independent lines of study over the years. As we have just seen, it is a natural extension of inflational cosmology. As Guth put it, "What happened before inflation? More inflation". The 'pre-inflation' part of our Big Bang was what we now call the multiverse. Andrei Linde came up with the idea of eternal chaotic inflation in the early 1980s, and there have been other variants. Quantum fluctuations continually generate inflating regions, distinct universes which are forever separate from each other and from ours. As mentioned in the last chapter, parallel universes were invoked as early as the 1950s to explain some of the paradoxes of quantum mechanics. According to this 'Many Worlds' interpretation, everything that quantum mechanics predicts could happen does in fact happen, in a forest of parallel universes. In M-theory violent collisions are envisioned between branes in a highly multi-dimensional space that produce new universes. It has been suggested that other universes may exist separately from ours in extra dimensions – even that the 10^{500} possible solutions to string theory may each be realized in different universes of the multiverse. Lee Smolin has proposed that new and totally independent universes can be created in the depths of black holes – every black hole may contain the seed of a new universe. Alan Guth and Edward Harrison have even considered the possibility that 'artificial' universes could be made by imploding material to form black holes. Finally, as elaborated in the next chapter, the concept of a multiverse has been invoked to explain why our universe appears to be 'fine-tuned' for life. In all these scenarios, our entire universe is just one small part of a much larger ensemble: the multiverse.

Each universe in the multiverse would have formed with different 'initial conditions', and would therefore have its own unique properties. These could conceivably vary enormously and fundamentally from one universe to another. The multiverse allows for any kind of universe you can imagine, and probably more (it has been said that, in the cosmos, "anything that is not strictly prohibited is absolutely mandatory"!). The very laws of physics may differ. Depending on the strength of gravity, some universes might immediately recollapse while others may expand so fast that no structures such as galaxies or stars could ever form. In some cases the laws of physics may not allow normal (atomic) matter to form, and inert dark matter may dominate. On the other hand, could there be super-laws of physics that apply to the entire multiverse? If mathematics were based on an ethereal Platonic world of ideals, wouldn't these have to apply everywhere? We live in a universe with three dimensions of space and one of time – could there be different numbers of dimensions in other universes? All of the 'coincidences' listed in the next chapter as favouring the existence of life in our universe could have been different in other universes, with radically different consequences.

There are even more bizarre possibilities. Lee Smolin conjectures that the universes in a multiverse could exhibit heredity and selection, if they are formed in black holes. Selection could arise if a given universe was exceptionally successful in producing black holes, and through them, new universes. If artificial universes could be created, what about 'designer universes'? Is our entire universe a grand experiment, or perhaps a computer simulation?

Where does all this leave our own Big Bang universe? Of course its properties (space and time dimensions, physical laws, etc.) remain unchanged. Einstein's theory of general relativity still describes its geometry and evolution. We understand the physics of our universe back to 10^{-10} seconds after the Big Bang, and the LHC will take this even further back to 10^{-15} seconds. However, the 'instant' of the Big Bang itself may now be explained by a quantum fluctuation in a sea of false vacuum starting the process of inflation. If so, a pre-existing cosmos must have existed. And if it can create our universe, then it can create others. Thus the multiverse.

According to the multiverse scenario there *is* an 'outside' (beyond our universe), and there *was* a 'before', although we can probably never directly know about either. Our universe is just one of a vast number of disjoint universes, and the Big Bang of our universe was just one event in a vast, perhaps infinite, multiverse. Our universe may share a 'common ancestor' with others. Interestingly, inflationary cosmology is somewhat reminiscent of the steady state cosmology of the 1950s, which was promoted by Fred Hoyle and colleagues as a possible alternative to the Big Bang theory, except that it would now pertain to the entire multiverse rather than just our universe.

The mind boggles at the speculation let loose by the concept of the multiverse. Needless to say, there is no evidence one way or the other – but there may be someday. For example, one competitor to inflationary cosmology is a 'cyclic' cosmology advanced by Paul Steinhardt and Neil Turok. They propose that we are living in a three-brane (see Chap. 4) that violently collides every trillion years or so with another nearby, parallel three-brane. The 'bang' from the collision (dubbed the 'big splat') initiates each new cosmological cycle in a 'big bounce'. There is no beginning of time in this eternal, cyclic universe. Primordial gravitational waves are predicted by standard inflationary cosmology, but not by the cyclic model. The gravitational waves would leave an imprint on the polarization of the microwave background radiation, so polarization measurements using the latest satellites may give the first answer. It is not inconceivable that other kinds of interactions of our universe with others could someday provide observational evidence for their existence. As we have seen over the past several decades, cosmology is no longer the science without evidence; spectacular advances in observations have been made, confirming a 'Big Bang' theory that seemed wild half a century ago. Who knows how future observations will confront the cascade of theoretical speculations being made today?

6. Is Our Universe Fine-Tuned for Life?

Coincidences?

It has been known for some time that there seem to be a number of 'coincidences' in the large scale properties and physics of the universe which appear to be essential for life as we know it.

These coincidences have been much discussed over the years, and for some they lead to an interesting conclusion. Several of the (somewhat overlapping) commonly quoted coincidences are summarized below.

We happen to live at a fortunate time in the history of the universe. Much earlier the heavy elements on which we depend (such as carbon and oxygen) did not yet exist; much later most stars will be too old to provide us with the energy and stability we need. Another way of expressing this is that the universe has to be just as big as it is for us to exist.

The balance between dark matter and dark energy is critical in the universe. If there were far more matter the universe would have re-collapsed before life could have evolved. If there were far less matter or far more dark energy, the universe would have expanded too rapidly for stars to form and for life to have had a chance. The expansion rate of our universe had to be in just the right range.

The irregularities in the very early universe could be neither much larger nor much smaller than they were. If they were much larger all structure would have collapsed into massive black holes, and if they were much smaller no structure (stars, galaxies) would have formed.

Even the fact that our universe has three dimensions of space and one of time is crucial for the existence of life as we know it.

P. Shaver, *Cosmic Heritage*, DOI 10.1007/978-3-642-20261-2_6,
© Springer-Verlag Berlin Heidelberg 2011

Planetary orbits are only stable in a universe with three spatial dimensions (not two or four), and electromagnetism only works in a universe with three dimensions of space and one of time. Universes with other dimensionality are conceivable, but they would not be able to support life as we know it.

Gravity is an extremely weak force. It is by far the weakest of the four forces, weaker than the electromagnetic force by a factor of some trillions of trillions of trillions. However, it is additive and long-range, and this gives it huge power across the entire universe. All the atoms of the Earth are working together through the force of gravity to keep you from drifting off into space. By contrast, the electromagnetic force has positive and negative charges which largely cancel each other out on large scales, and the strong and weak forces are short-range. So gravity, in spite of its relative intrinsic weakness, is actually the dominant large-scale force in the universe. It determines the curvature of space, the clustering of galaxies and the sizes of stars and planets.

If gravity were much stronger than it is, it would be far quicker in attracting matter to form galaxies and stars, and would produce far higher densities. Nucleosynthesis would start earlier and in smaller stars, which would have far shorter lifetimes. There would be no time for life to evolve. On the other hand, if gravity were too weak, no stars at all would have formed. Likewise, it is important that the mass-energy of our universe is very close (if not exactly identical) to the critical value. If it had started slightly off to one side or the other, the universe would have either collapsed or expanded too quickly for life to have developed. If the mass-energy of the universe is exactly equal to the critical density, it will continue that way forever; but a small discrepancy one way or the other would lead to a runaway, unfavourable for life.

If the strong force were much stronger, nuclear reactions would be so efficient that they would convert almost all the hydrogen into heavier elements. Without hydrogen there would be no life as we know it. If the strong force were much weaker, heavy nuclei would not be able to form. If the electromagnetic force were much stronger, electrons would be too tightly bound to their nuclei, and chemistry would not be possible. If it were much weaker, the electrons would be free, and again no chemistry would be possible. Thus, the ratio of the electromagnetic and strong forces had to be in the right range for the production of the elements.

If the weak force were much stronger, neutrons would decay and heavy elements would not exist. If it were much weaker, the hydrogen would be depleted by conversion into other elements. Furthermore, the weak force is important in generating the explosion of a supernova outburst, and disseminating the heavy elements that are necessary for life.

The mass of the helium nucleus is only 99.3% of the combined mass of the four hydrogen atoms that made it. The remaining 0.7% is converted into energy (a 'nuclear efficiency' of 0.007). If the nuclear efficiency were a few percent bigger or smaller, matter would either be 0% hydrogen or 100% hydrogen. In either case, there would be no life as we know it. The neutron-to-proton mass ratio is also important. The neutron is about 0.1% heavier than the proton. If it were even a fraction of a percent less massive than the proton, the protons would decay, and there would be no nuclei and no chemistry.

Neutrinos were important in the production of the primordial elements in the early universe, and they are important today in the dispersal of heavy elements from supernova explosions. In the former case, neutrinos could reduce the number of neutrons; if this process were too effective there would be less helium produced. In the latter case, the degree of coupling between neutrinos and atoms can seriously affect the outflow of heavy elements into the interstellar medium.

When scientists were first trying to understand stellar nucleosynthesis, there seemed to be a bottleneck in going from helium to carbon. And if carbon couldn't be made, then neither could any of the heavier elements – or life itself. Only the primordial elements would exist, and the complex chemistry of life would be impossible. However, as carbon and life clearly do exist in our universe, the stars must have some way of making carbon from helium.

The problem was that, while two helium nuclei can fuse together to make a beryllium nucleus, that beryllium nucleus is too unstable to fuse with another helium nucleus to form a carbon nucleus. Fred Hoyle grappled with this problem and came up with an idea. He suggested that the carbon nucleus might be able to exist in a so-called excited state that can resonate with the combined energy of the beryllium nucleus and a helium nucleus (excited states are like musical harmonics). If this were possible, the beryllium and helium nuclei would be able to form an excited carbon nucleus, which could then decay into the usual stable ground state of carbon. Hoyle computed the energy that this excited state would have to have, and encouraged experimental physicists at Caltech to search for it. He pointed out that, if the excited state were not found, they would only have wasted 2 weeks, but if it were found, they would be famous. They did find it, at almost exactly the value Hoyle had predicted. The nucleosynthesis bottleneck had been removed, and the road was now clear for the production of carbon and all heavier elements up to iron.

If the nuclear (strong) force were just a few percent different from what it is, the resonance wouldn't work, and our universe would be devoid of carbon, all heavier elements, and life. Furthermore, oxygen has a similar excited state; if it was just 6–7% greater it would resonate with carbon and helium nuclei, and the high efficiency of this process would quickly deplete all of the carbon; again our carbon based life would not exist. Hoyle's own reaction to all this was to say that it was too much of a coincidence – that it made the universe seem like a 'put-up job'. Many others have taken the view that this is just how it is; if it were different, we simply wouldn't be here to reflect on it.

In the 1930s Paul Dirac noted that the ratio of the strength of the electrical force to the gravitational force between an electron and a proton is a very large number: 10^{39}. As both forces obey the inverse square law, this is a constant (and therefore fundamental) number. He also noted that the size of the observable universe is (presently) about 10^{39} times the size of a proton. He found this to be too much of a coincidence. As the observable size of the universe is time-variable, this coincidence led him to suspect that the gravitational force may also be changing with time. However, observations of spacecraft in our solar system and of pulsars across the galaxy have shown no evidence for varying gravity. In 1961 Robert Dicke explained the coincidence in the following way. Stellar ages depend on the ratio of electrical and gravitational forces. Life depends on stars that have had time to mature but have not yet died. When stars reached this age, the universe had expanded to about 10^{39} times the size of a proton. Thus, Dicke argued, the similarity of the two large numbers is actually a requirement for life.

Finally, one more coincidence. As we saw above, the cosmological constant predicted by quantum mechanics (specifically the random vacuum fluctuations) is 10^{120} times greater than indicated by the observations of distant supernovae. At this enormous value no galaxies would ever have formed in the universe. Steven Weinberg came up with a solution to this problem several years before the supernova observations. Even if the positive and negative fluctuations in the vacuum are huge, they vary from place to place and time to time, so there must be some occasions when they by chance come very close to cancelling out. It is only such near-cancellation regions that could produce a universe that we could live in. We are here, so ours must be one of those. The cosmological constant in our universe does not have to be exactly zero, but it must be fairly close to zero. This is what Weinberg predicted years ago, and it is what the supernova work later found.

These and various other coincidences make some scientists feel that our universe appears to be 'fine-tuned' to produce life. So fine-tuned, in fact, that they feel it may call for a radical new approach to cosmology.

But *how* fine-tuned are these coincidences, exactly? How precise and significant are they? Can this be quantified? Many of them seem to cover a very large volume of parameter space. Should they really be called coincidences, or would words like 'consistencies' or 'compatibilities' be more appropriate? While many of them are certainly very broad indeed, a few are quite striking, such as those involving the strong, weak and electromagnetic forces. Hoyle's and Weinberg's predictions, subsequently confirmed by experimental and observational evidence, were impressive. All in all, an interesting set of coincidences has been identified, which, taken together, may be significant. If so, how do we explain them?

The Anthropic Principle

We can only exist in a universe that has made our existence possible. That is the essence of the anthropic principle, which has been stated in many ways. As observers, we shouldn't be surprised to see that the laws of nature and the properties of the universe are consistent with our existence. That is inevitable. It could not be otherwise, as we are certainly here. This realization goes back at least a hundred years, but the term 'anthropic principle' was coined by Brandon Carter in 1973. He later regretted his choice of the word anthropic, as it implied that the principle applied specifically to humans (why not seaweed, or rock?).

The version of the anthropic principle given above is called the weak anthropic principle. Many consider it to be just a tautology (a statement that is necessarily true by simple logic). A strong anthropic principle has also been postulated, according to which the universe *must* have the properties required for life to develop in it at some stage, although this is regarded to be speculative.

In any case, the fact that the universe appears to be 'just right for life' seems to some scientists to be sufficiently compelling as to require a special explanation, and the multiverse concept seems to fill the bill. If there is a multitude of universes, each with its own set of properties and physical laws, then a subset could exist which happens to have the conditions appropriate for life as we know it. Obviously, as we are undeniably here, our universe must be one of those. The large number of universes in a possible multiverse makes the probability of at least some of them having the properties necessary for life arbitrarily high.

There is much debate over the anthropic argument. For some, it provides a plausible answer. For others it (and the entire multiverse concept) seems unscientific if there can never be experimental or observational evidence of any of the other universes. And many scientists are hoping to find a unique Theory of Everything that will explain why our universe is the way it is without having to invoke other universes. For them, there is only one universe – ours.

7. The Universe on Fast Forward

For a bit of light relief, and in preparation for the next chapters, let's now go back to our 'own' parochial universe and 'run it forward' from the Big Bang to the present, in the process summarizing the preceding chapters. Here's what it would look like according to our current scientific knowledge.

The Violent First Minutes

From the instant of the Big Bang itself, the universe was exceedingly hot and dense; it was rapidly expanding, which caused it to start cooling and decreasing in density. It raced through several significant phases in just the first second. Our understanding of the physics involved in the earliest phases is speculative at best, but the physics at the end of the first second is well known to us.

The period from the Big Bang to the so-called Planck time, during which chaotic quantum fluctuations dominated, lasted for just 10^{-43} seconds (an astonishingly short one ten million trillion trillion trillionth of a second). The temperature was far too high for any of the forces and particles we know in physics to exist independently; they were all part of an immensely energetic 'cosmic soup'. At the end of this period, the temperature had dropped to 10^{32} K, enabling the force of gravity to 'freeze out' (a 'phase change', similar to that when steam cools and forms water), and become a distinct force.

There were then two forces in the universe: gravity and a unified force which was a combination of the strong, weak and electromagnetic forces. The temperature continued its rapid fall, and when it reached 10^{28} K, 10^{-36} seconds after the Big Bang, the strong force separated from the electroweak force. This is thought to have released an enormous amount of energy, causing inflation, a sudden expansion of the universe of phenomenal proportions, as described in Chap. 5. This extraordinary event is thought to have lasted from about 10^{-36} to 10^{-33} seconds, to have increased the size of the universe by an immense factor of at least 10^{30}, and to have produced the extreme smoothness of the universe on which were superimposed the small quantum fluctuations that were the seeds of structure formation.

At 10^{-10} seconds the temperature had dropped to 10^{15} K, cool enough for the weak force to freeze out and separate from the electromagnetic force. The four forces we know today, gravity and the strong, weak and electromagnetic forces, now existed independently. Meanwhile, the well-known elementary particles began to appear, first in frenzied interactions. By 10^{-3} seconds (a millisecond) the temperature was down to 10^{12} K, and protons and neutrons were able to exist as stable particles.

When the universe was a second old, its temperature had fallen to 10 billion degrees and nucleosynthesis could begin. By an age of several minutes the nuclei of hydrogen, helium and a few other light elements had formed, but the process ceased rapidly as the temperature of the universe cooled too fast for heavier elements to form (they had to wait until stars provided the right conditions for their formation, hundreds of millions of years later).

Recombination

For the next several hundred thousand years, the universe was dominated by the hydrogen and helium nuclei, electrons and photons. It was a hot plasma, opaque to electromagnetic radiation because the photons could not travel freely due to constant interactions.

After 380,000 years, when the temperature had dropped to about 3,000°, protons and electrons were finally able to (re)combine to form hydrogen atoms (the recombination epoch). The universe became neutral, and the photons were freed from the plasma to travel unimpeded throughout the universe. The universe suddenly went from being opaque to being transparent. This was a major phase change in the universe.

The photons freed at that time are actually the ones we see today as the 'cosmic microwave background' – they give us a precise snapshot of the universe as it was 380,000 years after the Big Bang. It appears to us as a 'wall', because the universe beyond it is opaque, while the universe between it and us is transparent. As described in Chap. 2, we can clearly discern the 'ripples' that originated as quantum fluctuations and that would much later become the structures that formed galaxies and stars.

The Dark Ages

As the temperature continued to fall the universe entered a period called the dark ages. There was no source of light, aside from the fading afterglow of the Big Bang. The universe was comprised of neutral atomic matter (the primordial elements hydrogen, deuterium, helium and lithium), and the mysterious cold dark matter (and presumably also dark energy). However, throughout this period the tiny residual clumpiness of the matter – the structure revealed in the microwave background – continued to increase in amplitude due to simple gravitational attraction. Even though the universe as a whole was expanding, the overdense regions of matter accreted more and more matter from the less dense surrounding regions, and continued to become denser and more massive through gravitational self-attraction. Slowly but surely, this process inexorably built up the concentrations of matter that would ultimately form stars and galaxies. It was an unspectacular but critical period in the history of the universe, and took some hundreds of millions of years. It was essential in setting the scene for the action to follow.

So far there have been no observations of this period. As most of the atomic matter at that time was hydrogen, it may eventually be possible to observe the epoch using the famous 21 cm line of neutral hydrogen at radio wavelengths. Alternatives based on lithium spectral lines have also been discussed. Considering the extraordinarily rapid development of astronomy over the past century, it wouldn't be too surprising if observations of the dark ages were made sometime in the coming decades.

Reionization

Reionization, the transition from a totally neutral universe to a highly ionized one due to the formation of the first stars, effectively took place when the universe was about 400–500 million years old (about 13.2–13.3 billion years ago) according to observations made with the WMAP spacecraft. It was another 'phase transition' of the entire universe, as most of the hydrogen went from being neutral to being ionized again (this time however, the universe remained transparent to most of the electromagnetic radiation, as its density had decreased so much that the photons were no longer trapped by the free electrons). Star and galaxy formation proceeded rapidly, and before the universe was a billion years old it was lit up and almost fully ionized.

In principle this reionization epoch should be detectable as an all-sky 'step' in the radio spectrum of the distant universe. Our expected view looking back at the dark ages is of a 'wall of neutral hydrogen', and this causes the spectral step. On finer scales we will see the neutral hydrogen fluctuations, statistically or individually, and, like the CMB fluctuations, these will convey a great deal of information about the large-scale properties of the universe. The possibility of observing the reionization epoch has become a hot topic in modern astronomy.

Theoreticians have also been working on the nature of the first stars, how simultaneously they formed across the universe, and how sudden or gradual the reionization epoch was. Were the most massive stars the first to form, and did they (as expected) form in the most massive overdensities? Or did much larger numbers of low mass stars form first, synchronously all across the universe? The first stars had to form out of the only material available to them – the light elements – and this would have made their formation more difficult than that of stars today. It is likely that only very massive stars could have formed under those early conditions. How efficiently did the first stars and galaxies enrich the intergalactic medium with heavy elements? Did massive black holes form before or after the first galaxies? How could supermassive black holes, containing the mass of billions of stars, have formed so early? Compared with the relative simplicity and purity of the very early universe, the physics involved after the first stars were born became quite complicated. One then had to deal with the effects of rotation, magnetic fields, shock waves, irregular densities and radiation fields, gravitational interactions, changing elemental abundances – virtually the whole gamut of physics. As the well-known astrophysicist Martin Rees once commented, the elegant simplicity of the very early universe had degenerated into 'mud wrestling'.

Observationally, the reionization epoch is coincidentally at the limit in two respects: (1) it is at the limit of what our most powerful telescopes can detect, and (2) it is at the 'edge of the universe', the point beyond which there are no further discrete, luminous objects (stars and galaxies) in the universe – it is the 'end' of classical astronomy, the last frontier. And a very exciting one. We have been able to map the evolution of the number of quasars and galaxies (their 'space densities') over cosmic history, and as we look back in time we can see the numbers plummeting as we approach the reionization epoch. We have found the near side of the transition. Looking a bit further back, we should be able to see the wall of neutral hydrogen increasing into the dark ages. The most distant quasar found so far existed when the universe was about 800 million years old, and it was already 'knee-deep' in neutral hydrogen; the most distant gamma-ray burst and the most distant galaxy known both existed even earlier, about 600 and 500 million years respectively after the Big Bang – just 4% of the present age of the universe. Studies of these most distant objects give telltale signs that they were immersed in some of the neutral hydrogen on the near side of the dark ages: the blue (short wavelength) ends of their spectra are abruptly cut off, as expected from absorption by intervening neutral hydrogen. We can see the universe rapidly 'closing in' as the more and more distant objects we find were increasingly immersed in the neutral hydrogen of the dark ages.

The next generation of large ground based telescopes (the Giant Magellan Telescope and the Thirty Meter Telescope, both under development in the U.S., and the 42-m European Extremely Large Telescope), and space telescopes (in particular the NASA-led James Webb Space Telescope) should be capable of seeing the very first stars and galaxies.

The Quasar Epoch

The reionization epoch was just the start. The growth of structures, stars and galaxies accelerated in pace all across the universe. The young galaxies began to interact violently with each other, creating still more star formation. Many galaxies merged with each other to form still more massive galaxies (when one galaxy was much larger than the other this process has sometimes been referred to facetiously as 'galactic cannibalism'). Matter was dragged to the centres of galaxies, producing and feeding supermassive black holes at their nuclei. Quasars, monster radio galaxies, and huge amounts of star formation were hallmarks of this phase in the history of the universe.

This phase became known as the quasar epoch, a period of intense activity in the history of the universe. Not surprisingly, it coincided with the peak in both the star formation rate and the incidence of gamma-ray bursts. The bulk of this exceptional cosmic activity occurred between about 2 and 4 billion years after the Big Bang. Like the phase transitions of the recombination and reionization epochs, this period was extremely well coordinated across the entire universe. The number of quasars and the rate of star formation fell off steeply both before and after this period. Maarten Schmidt, who discovered quasars and first mapped their evolution, called this the 'rise and fall of quasars'. The number of quasars at the peak of this epoch was a thousand times what it is today. The fact that the peak is so narrow shows that quasars do not have very long lives – certainly not as long as the lifetimes of most stars.

It was once suggested that the steep decline in the number of quasars beyond the peak (i.e. at earlier times) might be due merely to obscuration by dust in intervening galaxies or the intergalactic medium, rather than a true decline in the number of quasars. However, we now know that this is not the case. Quasars found by virtue of their radio emission (which is not obscured by dust) show the same decline.

The spectra of quasars at different distances from us are remarkably similar. Their heavy element abundances are difficult to determine with precision, but they seem to be greater than in many stars today. Most remarkable is that this is also true for the most distant quasars for which we have adequate spectra, which existed when the universe was less than a billion years old. Their spectra, like those of all quasars, show prominent features due to iron. How can iron – the end of the line in stellar nucleosynthesis – have been made and distributed so early in the history of star formation? This presumably indicates that the early stars that produced these elements were extremely massive. Only very massive stars have sufficiently short lifetimes, and only very massive stars produce substantial amounts of iron. This would be consistent with other arguments, mentioned above, that very massive stars were amongst the earliest to form.

The quasar epoch began to fade away as galaxies relaxed into more genteel orbits about each other, the rate of violent close encounters decreased, and the interstellar gaseous fuel that produced both the stars and the supermassive black holes started to become depleted. Motions within the galaxies themselves became more ordered and relaxed, so there was less of the chaos and interactions that had led to rapid star formation and inflows to the centres of the galaxies. Both the number of quasars and the star formation rate decreased rapidly with time following the peak of the quasar epoch, and have continued to decrease (albeit more slowly) to the present day.

The Buildup of the Elements

The abundances of the primordial elements were established in the first few minutes of the universe, and have remained almost unchanged ever since. By contrast, the abundances of the heavy elements have been increasing (relative to hydrogen) ever since the first stars formed over 13 billion years ago.

As discussed in Chap. 2, the heavy elements are all produced in stars. The first stars started with the pristine primordial elements (hydrogen, deuterium, helium and lithium), and initially contained no heavy elements whatsoever. But these first stars, which are thought to have been very massive, cooked up a cocktail of heavy elements, and spewed them out into the interstellar (and intergalactic) medium in a short time. Sophisticated computer simulations give increasingly detailed indications of how these first stars were formed, and how they released the newly created elements. One crucial question is how widely the new elements were dispersed; it is now thought that a large fraction of the volume of the universe at that time could have been enriched rather quickly by the first generations of stars.

The next generation of stars formed in the usual way out of the gas in the interstellar medium, which by then contained some heavy elements. These new stars in turn produced more heavy elements, and spewed them out into the interstellar medium, adding to the heavy elements already present. Stars have several ways of expelling the elements they consist of, such as releasing giant shells of matter during various phases, planetary nebulae, novae outbursts and supernova explosions (massive stars actually disperse most of their mass into the interstellar medium over the course of their lifetimes). And so it went: successive generations of stars took their turn in adding to the growing buildup of heavy elements.

From 0% before the first stars formed, the heavy elements now account for about 2% of the total mass of elements in the universe today. This may not sound like much, but considering that the universe is pretty big, it amounts to an enormous mass of matter.

Direct evidence for the buildup of heavy elements is seen both in stars and in the intergalactic medium. Some stars that we see today were born over 13 billion years ago. At that time the heavy element abundances were still very low, and there hadn't been many previous generations of stars to create and distribute the elements. Therefore, we would expect the oldest stars we see today to have very low heavy element abundances – and they do, as low as 0.1% of their mass. By contrast, young stars in our galaxy, born recently out of today's interstellar medium, have heavy element abundances of about 2%, the same as the interstellar medium itself. We can also witness the buildup of abundances by observing absorption lines due to intergalactic gas clouds in the spectra of background quasars. We can observe such clouds in this way over most of the history of the universe, and so we can see the buildup happening 'in front of our eyes'.

Dignified Middle Age

Now, at 13.7 billion years, the universe has reached a comfortable middle age. The frenzy of the reionization and quasar epochs has died away. By now star formation has been going on in the universe for about 13 billion years. Many of the stars formed in the early phases have themselves now reached middle age. Many others have long since died, while successive generations of still others have both come and gone. The chaotic galaxies of the quasar epoch have merged, matured and evolved into the graceful 'grand design' galaxies – spirals and ellipticals – that now dominate the cosmic landscape.

But star formation still goes on today, just at a more dignified, less frenzied pace. Most of it occurs in gas-rich spiral and irregular galaxies. Elliptical galaxies have used up most of their interstellar gas, and so have little ongoing star formation. Much of the star formation in our own galaxy takes place in a ring extending between about 10,000 and 20,000 light-years from the centre of our galaxy. By comparison, our solar system is located in a quiet neighbourhood 27,000 light-years from the centre.

Supermassive black hole monsters still lurk at the centres of most galaxies, but most of them are now starved of the diet of infalling matter they had been accustomed to, and, like the one at the centre of our own galaxy, just sit there regally in a quiescent state. From time to time a morsel of interstellar matter may come too close, perhaps due to a perturbation by a neighbouring galaxy, and the monster enjoys a brief but active period of feeding. But such events are increasingly rare.

Our Sun and solar system formed 4.6 billion years ago, long after the first stars formed more than 13 billion years ago. By the time our Sun formed, well over 8.4 billion years of star formation activity had already taken place in the universe. A vast number of planets must have already formed before our Earth did. Many of them may have borne life – billions of years before our Earth even existed. And over the 4 billion years that life was evolving here on Earth, countless stars and their planets continued to form all around us, in our galaxy and across the universe. All of this continues today. These timescales dwarf the 200,000-year history of our own species, the 6,000 years of recorded history, and the 100 years of modern technology.

This brings us to one of the very big questions – how and when did life form on Earth?

8. The Cradles of Life

The process of star and planet formation provides the most likely opportunity for life as we know it to take hold in the universe.

Life as we know it wouldn't form inside a star, and it wouldn't form in the emptiness of intergalactic space, as these are respectively too hot and too cold. One might think that the early stages of the reionization epoch may have been possible when there were large quantities of raw material (gas) everywhere. But there were no heavy elements at that time – a requirement for life as we know it. At the other extreme one might think that billions of years from now an old and sterile planet could still acquire life, long, long after it had formed. Panspermia (the dispersal of life) between planets would be a possibility, but that just transfers the problem to another planet. Or the planet could finally, belatedly, give birth to its own home-grown life, perhaps after migrating into the 'habitable zone' of its stellar system (the habitable zone is the optimal distance range for life from the parent star, as described in Chap. 17). But it would then not have the advantage of the rich conditions that existed when our Earth formed, including the remnants of the parent molecular cloud, the bombardment by molecule-bearing comets and asteroids, and the turmoil of volcanic eruptions and dramatic changes in its atmospheric environment. The conditions provided by the formation of a new star, with its molecular cloud, chaotic protoplanetary disk, infant planets and other debris undoubtedly gave life the best chance to take hold. Furthermore, if life as we know it can only exist on planets, and if it forms as quickly as it can, then new planets around new stars are the most likely candidates. In the case of life on Earth, we have evidence that it had already formed 3.5–3.8 billion years ago, not so very long after the Sun and Earth themselves formed 4.6 billion years ago. So, to understand how life

P. Shaver, *Cosmic Heritage*, DOI 10.1007/978-3-642-20261-2_8,
© Springer-Verlag Berlin Heidelberg 2011

emerges, it is important to understand the processes of star and planet formation.

As mentioned above, the first stars were born some 13 billion years ago, and star formation has been taking place in all galaxies ever since. Even now, several stars are born in our own galaxy every year. Fortunately, some regions of active star formation are quite close to us, such as the famous Orion Nebula at a distance of just 1,300 light-years. In such regions we can watch the star formation process up close, almost in 'real time'.

The Formation of Stars and Planets

The process begins with the contraction and collapse of unusually dense regions of the interstellar medium – the gas and dust that occupy the space between the stars. These dense regions are the most massive objects in our galaxy, several million times the mass of our Sun. They are called 'molecular clouds', and for good reason: they contain large quantities of a wide variety of molecules. The molecules can exist because of the high densities and low temperatures of the clouds. Their formation can be initiated by cosmic rays, and also by stellar ultraviolet radiation in regions that are accessible to it. Dust grains can play a role by serving as platforms on which chemical reactions can take place.

Gravity causes these clouds, which are denser than typical regions in the interstellar medium, to begin to shrink and become even more dense. In response there is an increased outward pressure caused by the density and temperature of the cloud. One might think that this would lead to a balance between gravity pushing inwards and thermal pressure pushing outwards, as it does in stars. However, in the early stages the thermal energy in the cloud can be dissipated by photons emitted by the molecules and dust; as long as the photons can escape the cloud, the cloud's temperature can remain low. Therefore gravity wins, and the cloud continues to collapse. There are other factors that can slow the collapse, including magnetic fields, turbulence and rotation. Magnetic fields can provide a sort of stiffness to the cloud, braking the infall. Turbulence can cause the cloud to fragment into smaller clouds, but these smaller clouds can individually continue to collapse on their own, resulting ultimately in the formation of clusters of stars. In any case, sufficiently massive clouds can continue to collapse.

A critical point is reached when the inner part of the cloud becomes so dense that the cooling photons can no longer escape. They become re-absorbed by the cloud before they can get out. The temperature rises, and the infall is slowed. The central part of the cloud is now glowing intensely due to its increasing temperature and density, and can be observed through its surrounding cocoon of obscuring dust and gas at infrared and radio wavelengths. It has become a protostar, a well-defined hot core that will become a star when it is hot enough for nuclear fusion to begin.

The rotation of the cloud initially causes it to bulge out and flatten slightly across the plane of rotation. As the cloud collapses the speed of rotation increases (just as a spinning skater's rotation speeds up when she pulls in her arms), especially in the inner regions close to the protostar. This, plus internal friction, causes the infalling matter to become more and more concentrated towards the plane of rotation. It becomes a spinning accretion disk around the protostar, in time becoming denser and thinner, especially near the protostar itself. The infalling matter swirls around this disk as it makes its way to the protostar.

At the same time prominent jets of material are ejected along the polar axes of the protostar. They help the process of star formation by transporting away much of the angular momentum along twisted magnetic fields. They are spectacular phenomena to observe.

The magical final step in the formation of the star takes place when the temperature in the core of the protostar reaches ten million degrees, great enough for nuclear fusion reactions to take place. A star has been born. The time interval between protostar and star can be as short as a million years for the most massive stars to a hundred million years for the least massive.

The radiation from a massive new star continues to dissipate the surrounding gas and dust, and clears the region around the star. The accretion disk remains in place, but becomes thinner and denser. The more diffuse outer regions of the original molecular cloud are dispersed and eventually disappear.

A dramatic HST image of compressed clouds containing dust and molecules, showing two pairs of 'protostellar jets' shooting in opposite directions from newly forming stars (Image credit: NASA, ESA, and M. Livio and the Hubble 20th Anniversary Team (STScI)).

The interesting action is now in the accretion disk itself. Dust particles collide (violently or gently), and sometimes stick to each other. The more this happens, the more that other particles stick to the same coagulations because of the increased gravitational attraction. These coagulations grow into a wide range of sizes. The increasing mass of the larger coagulations cause them to become roughly spherical, again due to gravity. The largest become planets. These can be so large that they can sweep up the gas and dust in their orbital zones, like giant vacuum cleaners. Their orbital zones are ultimately cleared out, leaving gaps in the circumstellar disk occupied only by the new planets.

However, for the most part chaos reigns across the entire disk. Coagulations of all sizes are continually colliding with each other, in some case breaking to pieces, and in other cases leading to still larger objects such as comets, asteroids and planets. The life of a newly formed planet is tenuous, its environment very hostile with constant collisions. This period, which can last for a few hundred million years, has been coined the period of 'heavy bombardment'.

But in this chaotic way, planets, moons, asteroids and comets are formed as common by-products of the star formation process. The so-called 'terrestrial planets' in the inner regions of our solar system are made of the relatively rare rocky and metallic seeds from the original nebula. The outer huge 'Jovian planets' came from ices and hydrogen and helium gas. Asteroids and comets are the small rocky and icy leftovers of the inner and outer solar system respectively.

Until 1995 the presence of planets around other stars (beyond our solar system) was just speculation, but now we have found hundreds of them. Soon it will be thousands. There can be no question that the process described above is generic, commonplace in the universe.

Organic Molecules and Amino Acids in Space

Did life get a head-start from molecules which were already present in the parent molecular cloud even before a planet was formed? Organic molecules are those most associated with life as we know it, particularly those containing carbon, and several amino acids are the building blocks of proteins. Were some of them present in space even before the solar system formed?

Over 130 different types of molecules, from small to large, and mostly organic, are now known to exist in molecular clouds. A simple and robust molecule, carbon monoxide (CO), is so widely distributed throughout the interstellar medium that it is, along with neutral hydrogen, a major tool in mapping the structure of our galaxy and others, including the most distant. Other simple molecules found in molecular clouds include molecular hydrogen (H_2), water (H_2O), and common table salt (NaCl). At the other extreme are large molecules containing up to thirteen atoms, such as methanol (CH_3OH), acetic acid (CH_3COOH), propanal (CH_3CH_2CHO), cyanodecapentayne ($HC_{10}CN$), and alcohol (ethanol: CH_3CH_2OH).

Even some of the 20 amino acids which are the building blocks of proteins are present in interstellar space. One of them, glycine (H_2NH_2CCOOH), has recently been found in the tail of a comet, and there is evidence for it also in molecular clouds. Other amino acids have been found in meteorites. So comets and meteorites are considered to have been likely sources of organic molecules, including amino acids, on the early Earth.

Such molecules may certainly have given a head start to life on newly formed planets like the young Earth. At the very least they show that the complex chemistry needed for life can exist even in the harsh environment of space, making it all the more likely that it can get a hold on the surface of a young planet. Given the abundance of organic molecules (including amino acids) in these star formation regions, it seems possible that the nascent planets were already primed for life.

A Nascent Planet

The history of the first half billion years of the Earth is of course not at all well known. The Sun and the solar system formed about 4.6 billion years ago. The early Earth would probably have grown rapidly by accretion, and may have reached its present size in just several million years. Over much of the first half billion years, the Earth would have suffered under the 'heavy bombardment'.

The dramatic images of the fragments of comet Shoemaker-Levy smashing into Jupiter in 1994 impressed on all of us what a heavy bombardment can be like. Some of the disturbances were as large as the Earth itself. The impact on Earth of one of the largest asteroids known today, with a diameter of about 400 km, would vaporize both the asteroid and a considerable part of the Earth, eject large amounts of material into space, produce very high global temperatures that could evaporate the oceans, and sterilize the surface of the Earth.

The Moon provides a wonderful history of the early bombardment suffered by the Earth, because it is so close and has virtually no internal activity and no atmosphere. Compared with 4.3 billion years ago, the rate of major impacts on the Moon had already dropped to half 4 billion years ago and to 5% 3.5 billion years ago, and it is well under one percent today. Still, even today, tens or hundreds of thousands of meteorites fall onto the Earth each year, although most of them are 'pea-sized'.

The early Earth would have been hot, both because of the bombardment and because of active convection and chemical differentiation all the way down to the hot core. The earliest times would have been chaotic: bombardment from space, volcanism, molten rock, and possibly no atmosphere until gasses escaped from the inner turmoil. A very inhospitable place. Another complication was that, after the initial chaos, the Earth may actually have become cold for some time and any oceans may have frozen over, as the Sun had not yet reached its full luminosity. The atmosphere contained no oxygen, and there was probably no liquid water. The early atmosphere was probably rich in CO_2 (as is the case today on both Mars and Venus). But, when significant amounts of liquid water started to appear, the 'carbon cycle' (in which carbon is continually exchanged between the major components of the overall environment, including the biosphere, the geosphere, the oceans and the atmosphere) would have become important. Much of the atmospheric CO_2 ends up in the oceans, where it combines with calcium to precipitate as calcium carbonate into limestone. The oldest known rocks are about 3.9 billion years old. The oldest known sandstones have been identified on the basis of some hardy mineral grains such as zircon, with ages of

4.1–4.3 billion years. Thus there was some continental crust by then, perhaps several small continents.

Chaos, bombardment, eruptions, a changing atmosphere, some land and some ocean. The early Earth was a complicated place.

It would at least have provided some kind of platform for the development of the first life, although certainly not a very easy one. But, even starting with some of the building blocks of life and a nascent home, how did the first living organisms actually form? To address this question, we first have to understand what life actually is, and that is the subject of the next two chapters.

9. The Astonishing Diversity of Life

What is Life?

Although life is all around us, and we believe we can identify it when we see it, a clear definition of 'life' still eludes us.

The main characteristics of life that are often mentioned are metabolism, replication and evolution. Metabolism involves the conversion of external energy into the activity and essential components of the living entity, and the elimination of waste products. Replication involves copying and heredity, and evolution follows from variations in the presence of a challenging environment.

Other (sometimes overlapping) terms and concepts often considered in the definition of life include autonomy, growth, organization and complexity, homeostasis (regulation of the internal state) and response to stimuli. Some alternative definitions concentrate on those properties required for evolution, some the flow of energy, and still others on information.

To illustrate the difficulties in defining life, examples have been given of inanimate phenomena which have some of the properties of life (such as fire, which consumes energy and grows), and of obvious life forms which do not satisfy some of the standard criteria for life (such as mules and most other animal hybrids produced by breeding – zorses, hebras, zonkeys, tiglons, ligers – which are not capable of self-reproduction). Viruses are "organisms at the edge of life": they contain DNA, but they achieve replication by using the machinery of living cells; should they be considered to be alive or not?

Of course standard genetic analysis gives a clear indication of the presence of ordinary life as we know it today. But a more general

P. Shaver, *Cosmic Heritage*, DOI 10.1007/978-3-642-20261-2_9,
© Springer-Verlag Berlin Heidelberg 2011

and unambiguous definition of life is of obvious importance in the search for possible new life forms both on Earth and beyond, and for deciding whether some future laboratory may have succeeded in truly creating 'life'. Life elsewhere in the universe may conceivably be quite different from what we know on Earth. For its ongoing search for life in the universe, NASA's working definition is "Life is a self-sustained chemical system capable of undergoing Darwinian evolution". How comprehensive is this likely to be? At the moment our definition of life is undoubtedly incomplete.

Life Everywhere

There is a mind-boggling diversity of life in our world. It is everywhere on the Earth, throughout the seas, and can extend down kilometres beneath the surface of the Earth. Darwin, as usual, had a wonderful phrase for it: "Endless forms most beautiful". Just a pinch of Earth from the ground can contain billions of bacteria, along with decaying organic matter. Some years ago it was estimated that there may be as many as 100 million different species of life on Earth; recent estimates indicate that, including microbes, there could be upwards of a billion. Considering just some of the macro-organisms that we know in the world we can easily see around us, there are about 300,000 identified species of plants, 70,000 species of fungi, a million species of insects and 300,000 species of other animals, including 5,000 species of mammals of which we are just one. Species of micro-organisms such as bacteria are vastly more numerous. And these are just the numbers of species. The numbers of individuals are of course much greater: right now billions of billions of insects are alive around the world, and countless trillions of trillions of bacteria.

Counting species may be an impossible task, as they are forever changing and re-adapting in response to subtle (and not-so-subtle) changes in their environment and co-habitants. Life has managed to fill virtually every available niche, and close inspection shows that the forms of life found in each niche are extremely well adapted to it. Geographical latitude plays a major role in the abundance of life. Tropical rain forests, covering only a small fraction of the surface of the planet, probably contain more than half of the species on Earth. Coral reefs are the underwater equivalents. The numbers and diversity in both cases are spectacular. But looking on smaller scales reveals even more worlds of wonder. Just turn over a log or a leaf in a typical forest and you will immediately see many scurrying species, but there are many, many more there that are too small to see. An insect may carry a variety of species on its back without ever knowing. And we ourselves depend on trillions of bacteria that live and thrive in our digestive tracts (they actually outnumber our own cells!).

Over the past few decades we've discovered life in places we never expected. We have discovered extremophiles, organisms that survive in extreme environments of any kind. Some life forms (called thermophiles) thrive in water above the normal boiling point (100°C) surrounding the 'black smokers' at the ocean bottom, vents that energetically emit extremely hot mineral-rich water at the otherwise frigid depths of the ocean, where no life was supposed to exist at all. Some microbes live in rocks in the freezing cold valleys of the Antarctic. Then there are the lithophiles ('rock lovers'), that have been found as deep as a few kilometres beneath the surface of the Earth, in tiny water pockets within the rock. There are adaptations to the time dimension: endospores are cells that allow their parent organisms to become dormant, surviving for very long periods.

There are even extremes of extremophiles, known as polyextremophiles, in particular the tardigrades. These are microscopic, eight-legged animals. They can be found under solid ice, on Himalayan mountain tops, in hot springs and in the deep ocean. When necessary they can go into a dehydrated state of suspended animation for up to a decade, during which time their water content can be just a hundredth of normal and their metabolic rate less than a ten thousandth of normal. They can survive temperatures almost as low as absolute zero and as high as 150°C, and levels of radiation a thousand times more than we can. They can even survive the vacuum of space for some time (at least 10 days, as known from direct space experiments), as well as pressures greater than those of the deepest oceans. Scientists have yet to find any place on Earth cooler than 150°C without life. Even some ocean-floor vents hotter than 200°C contain biofilms. The deepest mud ever recovered – more than 4,500 m below sea level and a further

1,600 m down in the mud – contains microbes. All of these extreme life forms of course have major implications for the search for exterrestrial life, which will be discussed in Chap. 17.

The oceans can accommodate an enormous amount and diversity of life. They cover 71% of Earth's surface, and ocean trenches can be as deep as 11,000 m (deeper than Everest is high). The total volume in this three-dimensional world is huge. A remarkable "Census of Marine Life" took place in the years 2000–2010. It involved over 2,700 scientists from more than 80 nations. The amount and diversity of life found was staggering. About 250,000 marine species are now known. Microbes may not individually be so obvious and easy to identify, but as a group they completely dominate. There can be more than a billion microorganisms in just a litre of seawater or a gram of seabed mud. The total number of species of marine microbes may be upwards of a billion.

Another way of appreciating the world of life is to consider the total weight (or biomass) in different places and categories. Simple single-celled microorganisms such as bacteria may account for most of the total biomass on Earth. The biomass underground may exceed that on the surface. The total biomass of ants may amount to 10–20% of the entire animal biomass.

Energy is essential for life. On land, life depends on energy from the Sun acquired by plants through the process of photosynthesis, which both produces energy-rich molecules and 'fixes' carbon into organic compounds. In this way, plants provide the existential basis for all land-based species. Through the well-known 'food chain', herbivores eat the plants, small carnivores eat the herbivores, and large carnivores at the top of the food chain eat all those below. Near the surface of the seas and waterways photosynthesis also provides the energy. In this case the process is carried out by microscopic single-celled organisms such as phytoplankton and cyanobacteria. These are eaten by small creatures called zooplankton, which in turn are eaten by larger fish and mammals on and on up the food chain, culminating in sharks, whales and the like. The photosynthetic organelles of plants and algae evolved from primitive cyanobacteria. Photosynthesis gave rise to virtually all of the oxygen in the atmosphere of the Earth. Aside from photosynthesis, the only other source of energy for living systems comes from the inorganic chemistry along the mid-oceanic rifts and in rocks.

From this point of view, the world of life is divided into autotrophs and heterotrophs. There are just two types of autotrophs. Plants, phytoplankton and cyanobacteria are photoautotrophs: they carry out photosynthesis to acquire energy and produce organic compounds that are subsequently used as nutrition by other forms of life. Chemoautotrophs, on the other hand, derive their energy and synthesize organic compounds from chemical reactions, mostly in the environments of deep sea vents. Autotrophs are the 'primary producers' in the world of life (they are the good guys, as they get their energy and carbon compounds directly from the source). Heterotrophs are the 'consumers' (herbivores, carnivores and scavengers). That includes us. Heterotrophs eat autotrophs and other heterotrophs to obtain the energy and organic compounds necessary for life. This simple overview, based ultimately on the need for energy, explains much of the aggression in the world of life. It can certainly be, as Tennyson famously put it, "red in tooth and claw".

In this enormous world of life, it is not hard to think of examples of diverse forms ranging from the merely exotic to the totally bizarre. Here are just a few.

Bacteria comprise one of the three main branches of life. They are microscopic, enormously numerous and varied, and are present in virtually every possible environment on the planet, including extreme environments and the digestive tracts of animals. They multiply by cell division, in some cases every twenty minutes or so. Many bacteria can move themselves using a sort of propeller, and grab onto things with tiny filaments. Some bacteria employ photosynthesis, others get energy from inorganic chemicals, and still others are heterotrophs. Some bacteria can navigate using the Earth's magnetic field. Others can ride out rough times in reduced forms, resisting heat, radiation and acids.

Amoebae have remarkable properties for simple single-celled organisms. An amoeba might seem unimpressive, as it is a tiny microscopic blob with a constantly changing irregular shape. It slowly oozes its way around its environment in the soil, finding and feeding on bacteria and other organic matter. It engulfs them with what are effectively extensions of its body, so-called pseudopods, digests them and moves on. It reproduces every few hours by cell division, resulting in vast numbers. When the local food supply runs out, thousands to millions of amoebae aggregate into a multicellular unit: a cellular slime mould, or slug. The slug explores and searches for the top of the soil, guided by extreme sensitivity to light, temperature and chemical gradients. Once there the slug projects a long stalk topped by spores. The amoebae whose bodies comprise the stalk have sacrificed themselves for the spores which produce the next generation. The spores are brushed off and carried by passing insects to remote places, resulting in dispersal for the species. The newly emerging amoebae complete the life cycle.

Sponges are simple marine animals, but they have also been very successful. They have been around for over half a billion years, and individual sponges can live for up to 200 years. They live attached to the seafloor, anywhere from shallow shorelines to depths approaching 9,000 m. They range in size from millimetres to over a metre. They exist by the movement of water through them, driven by the motions of flagella (long, thin appendages); they feed on the nutrients (such as bacteria) carried along with the water. Sponges have no nervous system, but they do contain some of the precursors of neurons. They have cells that can change into other cell types, and they can remould their bodies. They can even regenerate if parts of them are broken off. Most sponges are hermaphrodites, producing both eggs and sperm.

One animal group that has thrived with and contributed to the rise of flowering plants are the insects. Some insect societies are so large and coherent that they are referred to as 'superorganisms'. Ants certainly fall into this category. Large ant colonies contain millions of individuals, working together and cooperating almost as a single entity. These colonies are organized in what is effectively a caste system, from queens to large-headed soldiers to small-headed workers. A female ant can become any of these, depending on the food and chemical stimuli she receives as a larva. Thus there is a division of labour. There is also effective communication (through a highly developed sense of smell) and an ability to cope with problems and situations that the colony may face from time to time. In some cases ants appear to be able to learn through teaching, an ability often thought to be unique to mammals. Some species raid the colonies of others, taking slaves in the process. Some even have 'farm animals'. A famous example is that of the aphids (plant lice), which ant colonies protect while the aphids are feeding and then 'milk' for the honeydew the aphids release. The ants obviously treasure the aphids, storing aphid eggs in their nests over the winter and taking them to newly formed colonies. Another example is a type of caterpillar, taken care of for the same reason. Yet other ants farm mushrooms. The superorganisms never cease to amaze.

Bats are the only flying mammals. Their ability to fly is based on thin membranes spread out between their exceptionally long and thin digits, and their flight is more precise and responsive than that of birds. The truly outstanding characteristic of the insectivorous bats is that they navigate by echolocation: they emit high pitch sounds to produce echoes, and in total darkness can identify and pluck insects from the air. At appropriate times of the year millions of bats emerge nightly from caves in Texas to attack migrating moths at altitudes as high as a few thousand metres. Most bats eat insects, and most of the others eat fruit. A few of the one thousand species, including the notorious Vampire Bat, live on a diet of blood.

Flowering plants comprise almost 90% of all species of plants, which is an indication of how successful they have been on the planet. They range from as small as a millimetre to 100 m tall Eucalyptus trees, and include cacti, rose bushes, oak trees, coffee plants, dandelions, tulips, coconut palms, corn plants and cabbages, to name but a few of the 260,000 species. Their huge success is based largely on the flower itself. A flower contains both the male and female components for reproduction: the stamen, which contains the pollen, and the carpel, which contains the eggs. Insects are attracted to the flower, and end up carrying pollen to other flowers, closing the reproductive loop through fertilization and the production of seeds. When you eat the fruit of a flowering plant, you are eating the ovary. Some plants are carnivorous, trapping and consuming insects to obtain nutrients. The Venus Flytrap is a particularly nasty example with a lovely name.

The life we find on our planet is truly astonishing: the sheer numbers (both of individuals and of species), the fantastic diversity in form and behaviour that goes beyond comprehension, and the almost unlimited range of habitats.

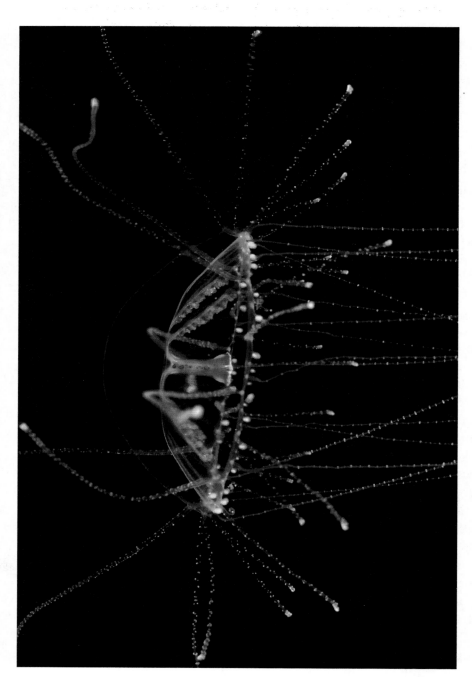

The family of life includes some spectacular species, such as this luminescent jellyfish from the Great Barrier Reef (Photo credit: Gary Cranitch, Queensland Museum).

10. The Astonishing Unity of Life

And yet – it is all just one single family of life. Underlying it all there is a genetic basis, which is common to all known life forms, from bacteria to humans. This has now been proven by the science of genetics.

The Code of Life

Life is based on a code. What we inherit from our parents and pass on to our children is a code, containing information. The atoms and molecules of which we are made may come and go (98% of the atoms in our bodies are replaced each year), but our genetic code remains with us forever. It is our genetic code that defines who we are and gives us the remarkable continuity and sense of self throughout life. And this basis of life is common to all life as we know it.

The well-developed field of genetics explains in detail how this works. The genetic information – the information of life – is encoded in the famous DNA molecule, whose double helix structure was discovered in 1953 by James Watson and Francis Crick. The double helix is like a spiral staircase, spiral ladder or twisted zipper, with many turns. Each 'rung' is made up of two molecules, called bases, joined together in the middle of the rung and each joined to one of the two sides (or 'strands') of the ladder, forming a so-called base pair. There are only four such molecules in total, and they are denoted by the letters A (for adenine), C (for cytosine), G (for guanine) and T (for thymine). The two bases comprising each rung can only join up in one of the following two combinations: A with T, and C with G. The genetic code is written as the sequence of bases along the length of one of the strands. Three bases in a row is called a codon, and this is the basis of the entire

P. Shaver, *Cosmic Heritage*, DOI 10.1007/978-3-642-20261-2_10,
© Springer-Verlag Berlin Heidelberg 2011

genetic code. A gene is a long continuous sequence of thousands of bases, and the DNA molecule contains thousands of genes. All the DNA, in its entirety, is called the genome. Virtually every cell in our bodies contains a complete copy of our genetic code.

The structure of DNA is ideal for all three of its functions: telling cells what they have to be and do, providing the basis for cell reproduction (growth), and for passing the genetic code on from generation to generation. The DNA molecule can be split in two along its length ("unzipped") by cutting each rung in the middle, thereby separating the two bases of each rung. The DNA molecule can then be reassembled, but the bases attached to each strand will have to match. So, if the bases attached to one strand have the sequence CTACGGATA, the sequence of the bases on the other strand has to be GATGCCTAT. One of these contains the original genetic code, and the other contains an exact (negative) template.

The instructions for cell activity are conveyed by the process of transcription. A length of the DNA is unzipped, and the genetic code contained in that section is copied onto RNA, a similar but one-stranded molecule. The RNA carries this section of the genetic code to special proteins in the cell, which in turn make the amino acids and proteins specified in the code. For replication, again the DNA is unzipped, and exact copies of each side are made; this duplication results in a new cell with a complete copy of the genetic code.

In the case of sexual reproduction, unzipped single strands of DNA from each of the parents are lined up side by side and zipped together with each other. The genes match up along the strands, although – crucially – a given gene can have different versions (called alleles) with different effects. Humans have 23 pairs of independent strands of DNA, and in reproduction each of the 23 strands matches up randomly and independently with any other similar strand. The number of possible combinations of strands is 2^{23}, or 8.4 million. To further mix things up, an amazing process called crossing-over produces an exchange of stretches of the strands, ensuring that the new strands in each pair contain both maternal and paternal DNA. Taken altogether, this is very much like shuffling a deck of cards, although a much, much larger deck and far more thoroughly. The DNA sequence maintains its original order, but the maternal and paternal genes (in their allele forms) are now mixed within each strand. This process explains how our children can be so different from us overall, while sometimes sharing certain specific characteristics with us and our grandparents: every individual is unique. More importantly, this process is crucial in producing the genetic variations that are required for natural selection, as discussed below.

All three of the functions mentioned above show that the exceptional structure of DNA plays a central role in the essential steps in life.

But how can the simple structure of DNA (and the genetic code itself) possibly provide for all the complexity of life? Not just human life, but all life. Well, for a start, a single codon, containing three positions, each of which can be occupied by one of four possible bases, allows for 64 possible combinations. That is more than enough to code for the 20 amino acids that comprise proteins. Then, a single human gene contains many thousands of codons, and the entire human genome contains about 21,000 genes and about three billion base pairs – equivalent to a library of thousands of books. Just in terms of the genetic code itself, there may be 10^{18} possible combinations of base pairs per species. The numbers are staggering. Furthermore, different genes often work together. The twenty amino acids can combine in a large number of ways to make individual proteins. And the proteins can interact with each other and with other genes, so there can be a huge multiplier effect in translating the genetic instructions into action. The whole process can be very complicated – it is sometimes referred to as a cascade.

There are also so-called epigenetic effects that can influence events and are stable over successive cell divisions but do not involve changes in the underlying DNA (the prefix 'epi-'signifies 'over' or 'above' – epigenetics is a system superimposed over the genome itself). It can play a major role in cellular differentiation, allowing cells to maintain different properties in spite of having the same DNA. Epigenetic characteristics are generally thought to be 'reset' in new generations of organisms, but they may sometimes be inherited and persist over a few generations. One of the major epigenetic mechanisms is 'DNA methylation', in which methyl groups bind themselves to various regions of the DNA and block the expression of some of the genes. The genome itself is unaffected, but some of its instructions are modified. This has sometimes been compared to a piano (the unchanging genome) being played by a pianist (the 'epigenome').

Breathtaking developments in technology are making rapid progress on the road to understanding many of these processes. The entire human genome was sequenced (decoded) a decade ago, and there are now many large-scale projects that go beyond. Genomic studies of many individuals give a knowledge of genetic variation in the population (the 'HapMap' project), the genomes of many other species have been sequenced (mostly living species, but some extinct), an effort is being made to map the entire suite of epigenetic marks, an Epitwin Project has been launched to study epigenetic effects in identical twins, an effort to construct a complete catalogue of human proteins is being considered, and there are many others.

The overall processes of life can certainly be complex, but even these complexities are shared by different species. Most important of all, however, is the astonishing fact that the underlying DNA code is amazingly simple, and common to all known life.

The Factory of Life

The cell is the fundamental unit of life. It is common to every living thing: all organisms are made of cells. There are single-celled organisms such as bacteria, and multi-celled organisms such as plants and animals. We ourselves are made of billions of cells, yet each of us began as just a single cell – the fertilized egg.

A cell is surrounded by a membrane which protects the contents of the cell from the outer environment. The beautiful thing about a cell membrane is that it 'just happens', by the lucky chance that the heads of some molecules love water while their tails hate it. In water such molecules spontaneously 'self-assemble', lining up together in a double layer in which their heads face the water and the tails, in the middle of the double layer, are sheltered from it. All cells have protective membranes. In addition some, such those of plants and many microorganisms, also have an outside wall surrounding the membrane, providing extra protection and structural integrity. The cell membranes and walls control what goes into and out of the cell in a variety of ways.

Entering a cell from the outside, one would be staggered by the frantic activity within. Millions of molecules are constantly moving about at high speed in the crowded interior. There are many types of independent bodies, called organelles, in our cells. The largest of these is the nucleus, which contains the DNA. It sits regally while its code is read and re-read by specific molecules, which then transfer the instructions at high speed to other molecules. Proteins, which carry out most of the functions, are forever being made and changing shape like contortionists. It may look like chaos, but everything has a well-defined function.

One of the vital functions of cells is making new cells. Most of the active cells in our bodies replicate themselves within about 24 hours. Much of this time is taken up with growth; the cell roughly doubles in size before cell division starts. Cell division results in two identical cells from the original one. All components of the mother cell have to be divided between the two daughter cells, and the DNA has to be replicated.

A full account of the workings of the cell leaves one with a sense of awe. And there are billions of cells within our bodies.

The Powerhouse of Life

Yet another ubiquitous feature of life is the mechanism for producing the life-sustaining energy that powers all biological processes. It is vital for all life-forms we know. It is no less than the basis of metabolism, one of the most important defining characteristics of life.

The process has to take place within the cells themselves; this is the only way that a sufficiently rapid supply of energy can be provided. So each cell contains tiny power stations, sometimes hundreds of them. The critical and universal process that takes place in these 'mitochondria', as they are known, is the final stage in the conversion of carbohydrates, fats, and proteins into carbon dioxide and water, resulting in the production of the essential energy. It is called the Krebs cycle, named after the biochemist Hans Krebs.

In a complex series of chemical reactions, the Krebs cycle ultimately produces a molecule, known as ATP (adenosine tri-phosphate), which is rich in energy. This molecule serves as the storehouse of energy, which can then be released as required for any function of the cell. It is known as the 'currency' of energy, as it is so widely and easily available throughout the cell. The energy release converts the ATP molecule into ADP (adenosine diphos-phate), which can readily be reconverted back into ATP with the input of further energy into the cycle. The Krebs cycle occurs thousands of times a second in a given cell; it has been said that the amount of ATP each of us processes in a single day is roughly equivalent to our total body weight.

The fact that this remarkable cycle, like the genetic code and the cell, is the same across the entire family of life – in bacteria, amoebae, plants and all animals including ourselves – has major implications for our understanding of life. The cycle must have arisen very early in the history of life, in order to be the same in all the different branches of the tree of life. Life as we know it must have arisen only once, from a common ancestor.

These three absolutely fundamental examples – the code of life, the cell and the powerhouse of life – empirically prove beyond any doubt that the family of life is one.

But of course there are other confirming lines of evidence from many other fields, such as molecular biology, comparative morphology, comparative anatomy and comparative physiology.

From molecular biology it is known that the same 20 amino acids are used in all organisms, whereas hundreds of others exist. The chiralities ('left-handedness' or 'right-handedness') of the molecular structures of DNA and amino acids are common to all of life. DNA is always right-handed, and amino acids are always left-handed. There are ubiquitous ion channels, present in the membranes of all biological cells; these are the controllable gateways into and out of cells, and they are vital for the life of the cell. Many crucial proteins are identical both in primitive life forms and in the most complex mammals. Common morphology, so obvious in closely related species, is the product of shared genetic elements. The entire genomes of all races of humans are 99.9% the same, and 98.5% the same as those of chimpanzees. The family of life is one, and we humans are as much a part of it as bacteria and amoebae. The unity of life is indeed astonishing – but it is a fact.

The 'tree of life' contains three main branches: the archaea, the bacteria, and the eukaryotes. Archaea and bacteria are prokaryotes: they have DNA but no nucleus. Eukaryotes have nuclei (which contain the DNA), and advanced eukaryotes are multicellular; these include algae, amoebae, fungi, plants and animals. Humans are represented by one tiny twig on the small sub-branch of mammals.

The title of the XXth International Congress of Genetics held in 2008 was "Genetics – Understanding Living Systems". For all topics – behaviour, ageing, cancer, biological clocks, epigenetics, parasitism, infection, mutations, neurology, ecology, cell biology and many more – examples were drawn and studies made from all forms of life, whichever was the most suitable and accessible in individual cases. The entire panorama of life was spread out as on a canvas, and lessons from one organism were applied to others. It is indeed the study of 'living systems' – ourselves very much included.

It is interesting to reflect that many years ago it was thought that there were two fundamental types of matter – living and non-living. The so-called 'life force' was supposedly what made the difference: substances that contained it were alive, and substances that didn't weren't. Now the life force concept has completely disappeared. Life is explained by the sciences of molecular biology and genetics. What was once called the life force has been reduced to the known laws of basic chemistry.

But – we still don't know how life started. . . .

11. How Did Life Begin?

The Origin of Life on Earth

While we now know how life works, we do not know how life started on Earth, and may never know for sure, even if we do succeed in creating life in the laboratory. The problem is that the historical trail may have been totally obliterated.

We do know the window of time for the formation of the first life on Earth. It could not have been earlier than 4.6 billion years ago, when the Sun and the solar system were formed. Evidence suggestive of life has been found in rocks that are 3.8 billion years old, and the earliest probable fossils appear in rocks 3.5 billion years old. So the window is from 4.6 to 3.8 (or 3.5) billion years ago.

As described in Chap. 8, the solar system at the time of formation was a swirling mass of gas and dust rotating in a disk around a young star (our Sun). Planets eventually formed from ever-larger coagulations of molecules, dust and rocks. The infant Earth was under continual bombardment by massive bodies of many sizes; the collisions would have made the surface of the Earth a most inhospitable place. Any form of life was probably not possible until about 4 billion years ago. In that case the window for the formation of life is reduced to a period of 200–500 million years.

How could life have formed? The first experiment to explore this question was performed by Stanley Miller and Harold Urey in 1953. They tried to re-create the possible conditions on the early Earth in a flask, and see what happened. Their choice of ingredients seemed plausible at the time: hydrogen, methane, ammonia and water (oxygen was excluded because its prevalence in the atmosphere is due to photosynthesis, which implied pre-existing life). They passed electrical discharges through this mixture to simulate lightning. After a week they analysed the resulting fluid,

P. Shaver, *Cosmic Heritage*, DOI 10.1007/978-3-642-20261-2_11,
© Springer-Verlag Berlin Heidelberg 2011

and found to their great delight that it contained amino acids, which are central to life as we know it. Their experiment was widely considered to be the first step on the road to creating life in the lab.

However the euphoria was premature, and it was eventually realized that there were serious problems. It is now thought that the ingredients used by Miller and Urey are probably not a very realistic representation of the early atmosphere. And, while the major triumph of the Miller-Urey experiment was to show how easily amino acids can be formed, it is far from trivial to assemble them into the macro-molecules of life. Back to the drawing board.

Looking back from the present, over the aeons when life has been present on Earth, we realize that we are faced with a major problem. The complex world of DNA/protein-based life goes back to the earliest life forms known. All life that we know (including the most basic microorganisms) is based on the same DNA/protein plan that we have today. This is the family of life described in the last chapter. Looking into the past, life as we know it existed, with the same basic elements as today, as far back as we can see. How was the leap possibly made from 'nothing to something' – from simple inorganic matter to life as we know it, with all the complexity of DNA and the protein world?

Could it have taken place by pure chance? Could the molecules have formed the complete, working complex structures of life as we know it today in a sudden chance event? The probability of this having occurred is virtually zero, even given the entire timescale of the universe. Fred Hoyle expressed this in his usual colourful way by comparing it to a tornado roaring through a junkyard and creating from the mayhem a fully operational 747 jumbo jet. Impossible.

Additional complications come from the 'chicken and egg' problem. The various components of life are interdependent – it is difficult to start with one without having the others already in place. DNA holds the code, but proteins are required to carry out the instructions. Which came first?

The protein-first case is supported by the ease with which some amino acids (the building blocks of proteins) can be created from inorganic material. If these can be assembled into proteins, if protective cell structures can be made, and if metabolism can result, a self-sustaining system may be produced. However, that system would not be capable of replication or evolution. On the other hand, while the DNA-first case is supported by the elegant simplicity of the double helix, this is not of much use if its genetic instructions cannot be implemented.

It has been suggested that RNA might provide the optimal starting point (the 'RNA world'). It is an intermediary between DNA and proteins. RNA can carry the genetic code, and, being single-stranded, it is simpler than DNA. In addition, RNA can also serve in the role of enzymes; perhaps it can also catalyse its own replication. Thus, it may be capable of being both the storehouse of the genetic code and the active agent that carries out the instructions, at least in simple life-forms. But one is then back to the original question of how such a system could emerge from inorganic matter in the first place.

Where might life have started? Darwin imagined a "warm little pond". Much thought has been given to the likely atmosphere of the primitive Earth, and how life might have started on the surface of the Earth in such an environment. Alternatively it may have started at the ocean bottom or even deep within the Earth, well shielded from the pandemonium that raged for hundreds of millions of years at the Earth's surface. The deep-sea vents that teem with life even today, and the extraordinary environments of extremophiles, which range from deep rock to glaciers, may give us an indication of how exotic the first life-forms and their environments may have been.

It is also conceivable that life on Earth originated somewhere else entirely. This is related to the hypothesis of panspermia: the possible dispersal of life across the solar system and perhaps beyond. Earth and Mars have been exchanging material for billions of years. Debris from an asteroid collision with one will have showered the other over the ages. If primitive life forms such as dormant bacterial endospores could survive the trip protected in a rock from the hostile environment of space, they could conceivably have populated their new destination. Perhaps we are all ultimately Martians! There have even been speculations about interstellar migration, unlikely as that may be. However, even if life on Earth originated somewhere else, this only shifts the problem of the ultimate formation of life to that other place.

Artificial Life?

Perhaps the answer to the origin of life will come from the actual production of a living organism in the laboratory. It is even conceivable that many avenues to creating life will eventually be found. This would of course be of absolutely fundamental significance. It would certainly solve the problem of the origin of life. However, even in that case, we may still never know which of the many avenues we find was the one that actually did start life on the early Earth, as the evidence may have been forever obliterated – although that will then be just a question of history rather than of fundamental science.

Some major steps towards artificial life have already been taken. The 'RNA world' has been particularly well studied, as it appears to hold much promise. The objective is to produce a minimal protocell comprising a protective envelope (cell membrane) enclosing a genetic polymer such as RNA (a polymer is a large molecule that is comprised of repeating subunits). An ability to absorb nutrients, grow and replicate would be required, and evolution would then presumably follow. This avenue is actively being pursued by several groups, most notably Jack Szostak and his colleagues at Harvard. It has already been established that, in appropriate environments, both the building blocks (nucleotides) of RNA and the cell membranes can form spontaneously. The assembly of nucleotides into strands of RNA is not straightforward, but it is enhanced in the presence of clay minerals, suggesting that clay-rich surfaces may have been important for the development of early life. So far, chains of 40–50 nucleotides have been produced in the lab (promising for a primitive protocell, if still a bit short of the billions of base pairs in some genomes today). An important step was the demonstration of the ability of RNA to catalyse the replication of other RNA strands ('self-replication'). Work on cell membranes in the lab has shown that they have the basic properties needed for the envisioned protocell. There have also been promising developments on cell division: in lab experiments some cells have been found to grow by extending thin filaments, which then break up to form new spherical cells. From studying these individual steps, the day may be drawing closer when they can be put together to produce the first living protocell.

Another approach starts with the well-known technique of transplanting genomic material. In the most common form of cloning, the DNA-containing nucleus of a somatic cell of an animal (i.e. a normal cell of the body, as opposed to those dedicated to reproduction) is removed from that cell and injected into an egg cell that has had its own nucleus removed. After test-tube incubation this altered egg cell is implanted into the uterus of an adult female, which eventually gives birth to an animal with the same genetic make-up as that from which the somatic cell was taken. Cloned animals so far include cats, cattle, dogs, horses, mules, sheep ("Dolly"), and rats. Primates (including humans) are more difficult for technical reasons, but it is likely that these problems can be overcome. Recently the genetic material of a fertilized human egg was successfully transplanted into another fertilized egg which had had its own genetic material removed. The reason for doing this was that the mitochondria in the original egg were damaged, and the transplant of the genetic material to the new cell (with healthy mitochondria) can avoid the severe complications that would otherwise have affected the new child.

In recent years it has become possible to transplant an entire genome from one species to another. In 2007 Craig Venter, Carole Lartigue and colleagues did just this. They extracted the genome of a cell of the bacterium *Mycoplasma mycoides* and inserted it into a cell of *Mycoplasma capricolum* (whose own genome had been removed). The researchers were able to confirm that the latter then had the same properties as the former. They had changed one bacterial species into another. It is true that these two species are close relatives of each other, and it remains to be determined how well the process will work on two species that are further apart, but in any case a crucial precedent had been set.

In a landmark development published in early 2010, Craig Venter, Dan Gibson and the team went even further. They made an entire genome from raw materials and inserted it into a cell, creating what they called a 'synthetic cell': a bacterial cell controlled by a synthetic genome. This cell metabolizes and replicates, like any normal living cell.

The entire genome, 1.08 million base pairs long, was assembled from four bottles of 'off the shelf' chemicals. The starting point was information from a digitized genome sequence of the bacterium *M. mycoides*. The synthetic genome is an exact copy of this genome, with the exception of some 'watermarks' that were added in non-coding regions in order to identify the synthetic genome (these stretches of DNA include coded passages from famous quotations and the names of the researchers). The genome was assembled in a series of stages, some of them with the help of yeast. A different bacterial species, *M. capricolum*, was selected to be the recipient of this new genome. Its own genome was removed, and the synthetic one inserted. The cell then began to function following the new instructions of the synthetic genome, making proteins and growing in ways characteristic of *M. mycoides* rather than *M. capricolum*. The cell had clearly been transformed from one species into another. It has since replicated over a billion times, each new generation containing a copy of the synthetic genome, including the telltale watermarks.

Of course this uses an already-existing cell (as Venter says, it makes sense to take advantage of 3.5 billion years of evolution!), but the cell is controlled entirely by the new synthetic genome. This opens the way to custom-made genomes, so that the living bacteria can be used for a variety of beneficial purposes such as producing biofuels and improving the environment (it also opens the way to potential misuse; scientists are well aware of this danger, and do everything they can to mitigate it).

In another important development also published in 2010, Michael Jewett and George Church announced the creation of synthetic ribosomes that can be used in a wide variety of applications. Ribosomes make the proteins specified by the DNA, and so are vital components in the cells of living systems. The RNA of the natural ribosomes of bacteria was removed, and replaced by RNA synthesized from chemicals. As in the case of the synthesized genome, this makes it possible to 'fine-tune' the ribosomes for specific purposes. It is another milestone towards making artificial life.

However, the creation of true artificial life will require that *all* components, not 'just' the genome and ribosomes, be designed and made ultimately from raw materials, and then assembled to make a complete cell. There are two fundamentally different approaches: top-down and bottom-up. In the former, one starts with a 'large' system (such as that produced by Venter and his team), and works downwards by eliminating unnecessary genes and understanding the others. In the latter one starts with a minimal cell that can be understood from the outset. Serious consideration has been given to the various steps and components that would be required to produce a 'minimal cell', the smallest cell that could be capable of replication and evolution. It has been estimated that such a cell may be possible with a genome containing only a couple of hundred genes – small enough to be understood before being built.

Radical alternatives to 'life as we know it' are also being considered. As mentioned above, our family of life may not be unique in the universe, and very different life forms could conceivably exist based on very different chemistry. Such alternatives could also be invented and created in a laboratory. One example, being pursued by Steven Benner and his team in Florida, involves a 12-letter genetic system rather than the four letters we know. It is said to be capable of satisfying many of the defining criteria for living systems, but it is not self-sustaining. Undoubtedly other alternatives will be explored over the coming years.

In spite of the impressive progress being made on all fronts, it seems that true artificial life may still be some time in coming. Perhaps that is just as well, considering the enormous societal and safety issues that remain to be worked out.

Why Life Anyway?

Why does life bother to exist? Why should a barren rock in space become covered with small, autonomous coagulations of matter that can move, grow and copy themselves? Why did molecules bother to come together in ways that would lead to life? What was the advantage? Why not just remain independent atoms and simple molecules (or inanimate rock) forever? Life is the thinnest veneer on the universe; if our present knowledge of the solar system is anything to go by, the living biomass may be considerably less than a thousandth of a trillionth of the total mass. It would seem to be an insignificant add-on. Why does it even exist?

Did life have a choice? Is it perhaps possible that, given the known properties and physical laws of the universe, life was actually inevitable? Life is as subject to the laws of physics and chemistry as anything else, so why shouldn't its origin be too?

Certainly protons and neutrons didn't have a choice – they were obliged by the laws of physics to combine to form atomic nuclei within the first minutes of the universe. They couldn't avoid it. Similarly, electrons were obliged to combine with nuclei to form neutral atoms 380,000 years after the Big Bang.

What about stars? Their origin was once also considered to be a mystery, but it is now considered to have been virtually inevitable. Gravity inexorably pulled matter together into smaller and smaller concentrations, until they were so dense and hot that nuclear reactions finally began and the first stars were born. Certainly the births of the first stars would have been more difficult than those of stars today. As mentioned in Chap. 8, the molecules formed by heavy elements provide important cooling, which assists the collapse into stars, but heavy elements did not exist when the first stars formed. This and several other factors affecting the formation of the first stars are still being discussed. They have a bearing on how fast the stars could form and what their masses would have been, but there is little doubt that they had to form. They 'self-assembled', in accordance with the straightforward laws of physics and chemistry. Nowadays we can observe stars forming in real time right in front of us.

If nuclei, atoms and stars were inevitable, then what about life? Its origin would undoubtedly have been much more complicated, but that doesn't mean it was impossible. Some of the perceived difficulties of forming life on Earth were elaborated at the beginning of this chapter. But modern approaches to the subject are more nuanced and sophisticated. Life probably didn't arise simply or suddenly. In the modern scenarios for the origin of life on Earth there is no 'spark of life' (this is a nice catchy phrase, but it's misleading). The origin of life was probably a comparatively unspectacular and gradual emergent process, and it would likely be difficult to nominate a specific time or event defining the 'start' of life. It most likely emerged out of a large number of complex and interacting intermediate steps, all of which were ultimately mandated by the preexisting laws of physics and chemistry. What was necessary to get life going was the simplest possible organism capable of metabolism and replication. From that basic start biological evolution would take over and lead ultimately to all the complexity we see in life today.

We know that complex organic molecules have been able to form in the harsh environment of space. The interstellar medium is (relatively speaking) rich in organic molecules. Many such molecules in space have been observed over the past several decades using radio telescopes that operate at millimetre wavelengths, and we are learning more and more about their formation. Such molecules are also found in icy comets, the remnants left over following the formation of our solar system from interstellar clouds.

Some of the building blocks of life are more easily formed than others. The formation of some amino acids is favoured by energy considerations – the energy of the system is reduced when they form. Some of the fundamental structures of life actually 'self assemble', such as cell membranes composed of molecules that love water at one end but hate it at the other. Most of the complex macromolecules of life require energy input for their formation, but both the Sun and the interior of the Earth provide copious supplies of energy. It is undoubtedly true that the vast majority of molecules formed have no relevance whatsoever to life. But, given that there were hundreds of millions of years available, if large molecules were constantly being formed, they would have had abundant opportunity to randomly 'explore' vast combinatorial possibilities and niches together with their neighbours. They could have eventually settled into the 'best' (often lowest-energy or most stable) configurations. How they would 'know' that they'd 'struck it rich' by becoming part of a macromolecule that was on its way to becoming part of life is not yet clear, but it may have had something to do with survival of the fittest amongst their peers. So their development into the precursor macromolecules of life may even seem to have been inevitable in retrospect.

Many possibilities have been considered. What more primitive intermediate stages might there have been? Pre-RNA replicating molecules? Primitive cells capable of capturing and using energy? One imaginative proposal is based on inorganic crystals. Crystals do grow, and, with inevitable impurities present, can 'evolve'. Could a rudimentary inorganic clay have served as the basis upon which the first life forms developed? Any evidence of such a precursor to life may long ago have disappeared, in the same way that there is no hint today of the wooden scaffolding that was used to make a still-existing Roman arch. Clays may well have assisted prebiotic chemistry in several ways. The difficulties that we see in various scenarios put forward so far may simply reflect our ignorance of possibilities that may have existed on the primordial Earth, but we are rapidly learning.

The question is whether such steps could, after countless generations, have led ultimately to the precursors of living systems. Many scientists think that life on Earth was indeed inevitable, but we don't know that, and we won't at least until we've been able to trace a line of plausible steps from prebiotic chemistry to life itself (or do it ourselves in a lab). Many have even conjectured that life was actually invented not just once but many times during the early history of the Earth, and that the one single family of life that we know was the one that finally won out in the grand competition.

The more we learn about prebiotic chemistry and possibilities for artificial life, the more we will know about the likelihood of each individual step. This will give us clues as to the most probable route(s) that emerging life on the early Earth may have taken, as well as any other forms of life that also obey exactly the same laws of physics and chemistry. It will give us an improved understanding of the overall likelihood of life of any kind having formed in the universe. The study of the prebiotic chemistry of the early Earth and the work towards artificial life are increasingly linked, and what is learned from one is relevant to the other.

Whether they were rare events or not, the processes leading to life on Earth were undoubtedly guided by the pre-existing laws of physics and chemistry in the universe. (This is the mundane flipside of the anthropic principle: rather than the universe having to be just right for us, we had to be just right for the universe.) In that case life would just as likely have formed elsewhere in the universe as here on Earth. Wherever the local conditions are suitable (an appropriate planet surrounding an appropriate star of an appropriate age), life would be possible. There could well be an abundance of life elsewhere in the universe.

In summary, one might say that there are three very different approaches to understanding how life originated on Earth. The first is the most straightforward: just find the evidence. But as the trail has probably been totally obliterated, this is likely impossible (although we may someday get clues from observations of distant planets that are just now being formed). The second is to do it ourselves: make artificial life. Then we know it's possible, and therefore no longer a great mystery. The third is to show convincingly that life was actually inevitable, given the laws of physics and chemistry in the universe.

At the moment we cannot say that we understand the origin of life, but the sophisticated studies now being done may eventually change that.

12. How Did Life Evolve?

Evolution Everywhere and all the Time

How did the huge diversity and complexity of life come about? While we do not yet know the origin of life, we do know how we got here from there: through the process of evolution, as proposed by Charles Darwin in his book *The Origin of Species* published in 1859. This is a simple but extremely powerful concept. Incredibly, this simple theory explains the extraordinary variety of life on Earth.

In essence, evolution requires just that populations of life forms (species) continually reproduce, that random changes appear in the genome from time to time, and that the environment is changing. The changes in the genome can occur through mutations due to a variety of causes, such as rare errors in copying the DNA and damage from radiation. As a result the effect of a gene may be altered. The result can be beneficial, neutral, or harmful, depending on the nature of the variation and on the environment. The environment is the final arbiter. It includes all aspects affecting the individual's life – food, predators, mates, temperature, sunlight, other species, geographical factors – everything. If the variation has a net positive effect in relation to one or more aspects of the environment, the individual will tend to thrive. If the variation has a negative effect, the individual will be at a disadvantage. In Darwin's time the cause of the putative variations was unknown, but now genetics has given the definitive answer.

The effects of these advantages and disadvantages play out on many scales. Within a given species, the individual will be more or less successful than its peers in that species. Ultimately, advantageous variations will tend to thrive in a species, and disadvantageous variations will tend to die out. In this way, the species as a whole will prosper, given a fixed environment. It will be at an advantage and grow relative to other species.

But the environment itself varies with time. Climatic conditions change, seas rise and fall, glaciers come and go, other species also evolve, grow and decline. A given species will only continue to thrive (and ultimately, to exist) if the variations occurring within its population can keep up and are ultimately beneficial in the face of the ever-changing environment. A sudden change in the environment – so sudden that evolution can't keep up, and life can't adjust quickly enough – can wipe out entire species, as we know from the extinction of the dinosaurs. Over 99% of all species that have ever lived have become extinct for one reason or another.

In the normal course of gradual evolution, entirely new species can be introduced. This can happen, for example, when there is a geographical split in a species, due to the migration of a subgroup. When there is long-term isolation between two groups, the characteristics of each group will continue to evolve, but now independently, and at some point the groups will become so different from each other that any sexual interactions between members of the two can no longer lead to reproduction. At this point the split between the two groups is irreversible, and they become two separate species. Further separate evolution can eventually make them entirely different in appearance and characteristics.

The slow process of natural evolution likely guarantees in most cases that the resulting development is only 'just enough'. Just enough to cope with the marginally new environmental circumstances. Drastic overshooting is undoubtedly rare – there aren't too many supersonic birds, and there are many animals that are decidedly cumbersome.

Of course, in general, evolution is not as neat and tidy as outlined above. This is easy to understand at the level of the genome. Each of us contains two genes (one from each parent) for a given characteristic (for example eye colour). In classical genetics one of them is 'dominant' and the other is 'recessive'; the dominant one determines the characteristic that will show up from that gene. The two versions of a gene are called alleles. This phenomenon was studied exhaustively by an Austrian monk, Gregor Mendel, at about the same time that Darwin's book was published. He chose to work with relatively simple garden peas, and he kept meticulous notes. He published his work in an obscure German-language journal in 1866, and its significance was not appreciated until the early 1900s. He effectively established the field of genetics, decades before others had even started down the same path. It is a great pity that he and Darwin had not been in touch, as Mendel's work provided the key to the variations that are the basis of Darwin's theory of natural selection.

Mendel was lucky in having chosen to study peas which were so simple genetically. Inheritance in most organisms is more complicated. In some, neither allele is completely dominant, and intermediate chacteristics are expressed. In other cases both alleles are expressed. It gets even more complicated. Most genes have more than just two alleles, in many cases several. And most genes are responsible, not just for one characteristic, but for several. Two or more genes can be involved in determining a given characteristic. Some genes can affect the expression of others. And finally, the nature of a characteristic can depend on the environment as well as the genes.

So the expression of the genome on the overall phenotype (an organism's observable chacteristics) can be very complex. But that does not in any way diminish the crucial importance of the genome; the genes are the units of inheritance, the genome carries the genetic code and the genes themselves throughout a lifetime, and it determines what will be inherited. The point, however, is that the phenotype, and the range of phenotypes in a given species, can be very rich. The totality of all the alleles for all the genes of all the individuals in a population is called the gene pool. It can obviously be vast and complex. The gene pool can be modified by mutations, random fluctuations within the gene pool (called genetic drift), non-random mating such as inbreeding, natural selection due to a changing environment, and new genes being introduced by sexual interactions with other populations (gene flow). The size and complexity of a gene pool makes it possible for rapid evolution to take place, even in the absence of mutations. The mutation rate, which can be very slow, does not limit the speed of evolution over the short term, which can be very rapid. However, mutations are still the ultimate source of the variations in the gene pool (each human is born with over a hundred new mutations). And only natural selection can consistently cause adaptive evolution.

The geography of the Earth has played a major role in the long-term evolution of life. So before we summarize the evolution of life, we should understand a few basic facts about the Earth, including its own evolution.

All the World's a Stage

The only part of the Earth we know directly is the outermost layer – the crust. That's our home, the stage for the pageant of life. It is extremely thin, typically less than 1% of the 6,400 km radius of the Earth (roughly equivalent to the relative thickness of an apple skin). We know indirectly about the inner layers – the mantle and the outer and inner cores. It is convection in the liquid outer core that gives the Earth its magnetic field. The rocks we see around us are basically of three types: igneous, sedimentary and metamorphic. Igneous rock originates from magma (molten rock) when it cools and solidifies, as we observe in flows from volcanoes. Sedimentary rock is formed when particles of matter accumulate, typically in layers in or near bodies of water, and over time become compressed into rock. Metamorphic rock started as one of the former but changed significantly, usually by deep burial and heating that changed its mineral makeup.

The fact that the shapes of today's continents on a world map seem to fit together like a giant jigsaw puzzle must have impressed many people, but it was the German meteorologist Alfred Wegener who first took it seriously and began to compare the flora and fauna of the continents. He found that they matched as would be expected if these continents had once been adjacent to one another. Studies of the seafloor in the 1950s and 1960s using echo sounders and magnetometers, and later satellite imaging, ultimately proved the validity of this hypothesis beyond any doubt. Vast ridges, deep trenches, large volcanoes and other extraordinary topological features were found, instead of the boring, featureless seabed that was expected. The mid-Atlantic ridge is 2.5 km higher than the seabed on either side, and is part of a vast system that extends into all the oceans. Most striking of all were the zones of reversing magnetic field directions, parallel to the ridge but radiating away symmetrically on both sides, like parallel zebra stripes on a huge scale – effectively a giant tape recorder. These symmetrical patterns occur everywhere along the ocean ridges. They are due to two major factors. First, magma is oozing out of the mid-oceanic ridges, spreading on both sides, and is magnetized by the Earth's magnetic field; this magnetization becomes frozen in when the magma solidifies. Second, the Earth's magnetic field has frequently reversed polarity over geological time. The patterns so revealed are stunningly clear and simple. There is no question: the seafloor is spreading, and this is clearly related to the movements of the continents away from each other. Such processes have been occurring for billions of years. Various terms for this global phenomenon are used: seafloor spreading, continental drift, and plate tectonics.

Thanks to developments such as the Global Positioning System (GPS) based on dozens of satellites, the movements of the continents can now be routinely measured in real time. North America and Europe are moving apart about as fast as your fingernails grow – several centimetres per year (that may seem slow, but it is fast enough for the Atlantic ocean to have formed in just 200 million years – very short in geological terms). The Earth's crust is moving all the time, and it is not merely drifting. The Chilean Earthquake of February 2010 suddenly moved an entire continent and shortened the length of a day! It shifted the city of Conception over 3 m to the west, and it even shifted the city of Buenos Aires, on the other side of the continent, by 4 cm. The effect on the length of a day was small (a millionth of a second), but easily measurable. The Japanese earthquake of March 2011 moved the main island of Honshu by more than 2 m. So continents don't just drift – they can also be moved suddenly and dramatically by major seismic events.

The plates in plate tectonics are pieces of the thin outer layer of the Earth. They move about the surface of the Earth as a result of the convective movement of the hot rock of the mantle below; the continents merely go along for the ride. The energy for this activity comes from deep within the Earth – both the heat remaining from the origin of the Earth, and the heat continuously generated by the radioactive decay of elements in the Earth. Almost all the seismic activity on the surface of the Earth happens at the interfaces of the plates: earthquakes, and volcanoes in particular. The famous 'ring of fire' around the Pacific ocean is a prominent example.

A wide range of phenomena result from plate tectonics. Some plates are colliding, subducted under others, causing great folds, enormous mountains and high plateaus. Some plates are in the process of rifting apart (the Great Rift Valley in East Africa, and the Red Sea). Some are sliding alongside others (the San Andreas fault in California) In some rare cases plates are drifting over fixed 'hot spots' in the mantle; these result in trails of volcanoes on otherwise unremarkable areas of seafloor such as the chain of the Hawaiian Islands, or on continents, such as the chain of the Yellowstone system of volcanoes and geysers. The Earth's seafloors are young, only about 200 million years old, while some parts of the continents are almost four billion years old.

Over time continent merging, breakup, and merging again would have taken place. By 1.6 bya, much of North America had accumulated into a continent called Laurentia. By 500–600 mya, many of today's continents had become joined together in one giant landmass, called Gondwanaland. By 200–300 mya, virtually all of today's continents had joined together into an even larger landmass, Pangea, which stretched from pole to pole. These changes would have caused great upheavals. The sea level was high during much of this period, possibly due to an exceptionally high system of mid-oceanic ridges, but eventually fell again, and the climate became dry. The movements in the Earth's crust never stopped; by some 70 mya Pangea was coming apart, and the distribution of continents looked more as it does today.

These continental movements had major effects on climate and biological evolution. As the land masses came together and separated, they closed off and then again opened passageways for currents in the oceans, which in turn had dramatic effects on climate, and therefore on life. Mountains, formed by the collisions of continents, affected atmospheric circulation. Regions that were arid and barren became wet and tropical, polar zones that had been oases became isolated and frigid. Continental movements can both create and destroy barriers to animal migration, profoundly affecting speciation. Of course we are more familiar with geographical features and their changes over relatively recent times.

The Isthmus of Panama formed just a few million years ago, closing off ocean currents between the Atlantic and the Pacific. The Himalayas created a high desert plateau in Tibet. The huge effects of the apparently trivial movements on the thin crust of the Earth cannot be exaggerated. Plate tectonics never stops.

Land masses are also necessary for the formation of glaciers, which have certainly had a huge impact on life. The oldest known glacial deposits are over two billion years old, probably due to a global ice age at that time. A number of prominent glacial periods took place about 600–800 mya. The most recent major glacial period took place only some 20,000 years ago, when ice covered almost a third of the land area of the Earth. Sea levels were obviously low at that time, and have since increased by almost 120 m. Another 60–70 m would be added if the ice caps on Antarctica and Geenland were to melt.

The most spectacular event we know of to have had a major impact on life was the asteroid impact that occurred 66 mya and wiped out the dinosaurs. The discovery was made in 1980 by Nobel laureate Louis Alvarez and his colleagues. What they found was tiny amounts of the rare element iridium in sediments that had been deposited exactly when the dinosaurs became extinct. While small in quantity, the amount they found was a hundred times that present immediately before or after the event. Iridium in the Earth is concentrated in the core; meteorites can have thousands of times the concentration of iridium that is found in the Earth's crust, so a narrow band of iridium spread widely over the Earth is a strong indication of a large meteoritic impact. The asteroid was estimated to have been at least 10 km in diameter. Such an impact would have wreaked total havoc. A global cloud of ejected dust would have darkened the planet, stopped the photosynthesis that plants and the entire food chain depend on, created shock waves that heated the atmosphere and precipated acid rain, produced huge tsunamis, and left the world prone to widespread wildfires. It's hard to imagine a greater disaster. The likely site of the impact is Chicxulub, in Yucatán, Mexico.

Aside from extraterrestrial impacts, many of the likely causes of mass extinctions of species were likely directly or indirectly due to changes in the land, the seas, and the atmosphere. Extinctions have been a part of life throughout the entire history of the Earth, and the list of possible agents is significant: climate change, changes in sea level (covering a range of 200 m over just the last 600 million years), ice ages, tectonic shifts and their effect on ocean currents and climate, volcanism and its effect on the atmosphere, and complications that many of these may have caused. The Earth is dynamic, and life has had to be nimble and adaptable to survive.

Measuring Age

How do we know the geological ages mentioned above? It's just basic physics. The major method for determining ages in geology is based on natural radioactivity. Radiometric dating (commonly referred to nowadays as 'carbon dating') goes back to 1905, when it was developed by Ernest Rutherford. Most chemical elements have several isotopes. All isotopes of a given element have essentially the same chemical behaviour, but each contains a different number of neutrons. Carbon-14 has two more neutrons than carbon-12, but both are still carbon. Certain isotopes, known as radioactive isotopes, are unstable. Radioactive decay involves a spontaneous loss of energy by the emission of ionizing particles and radiation, and results in the original atom being transformed into another type of atom. For example, carbon-14 is unstable, and transforms into nitrogen-14. The decay rate of a large population of carbon-14 atoms is a well-known number, and is characterized by the 'half-life' – the time elapsed when half of the population of carbon-14 atoms have transformed into nitrogen-14 atoms. The half -life in this case is 5,730 years.

It is easy to see why natural radioactivity is such an excellent geological clock. In the case of carbon-14, there is a continuing supply of the radioactive isotope. Carbon-14 atoms are continually being produced in the upper atmosphere by interactions initiated by cosmic rays. When plants fix atmospheric carbon dioxide into organic material through the process of photosynthesis, they incorporate an amount of carbon-14 corresponding to the fraction of that isotope in the atmosphere. When the plants die, they cease ingesting carbon-14, and the fraction they contain decays away exponentially in accordance with its well-known decay constant.

After 5,730 years, their remains will contain half of the fraction of carbon-14 found in present-day plants, after 11,460 years, one quarter, and so on. The carbon-14 radioactive clock is useful for ages up to about 60,000 years, when the amount of carbon-14 left is only a small fraction of a percent of the original. The subtleties of this method have been extremely well studied, and it is known to be highly reliable.

Many other radioactive clocks are also used, with a wide variety of half-lives. The overlapping results give us great confidence in the technique. Most of the unstable radioactive isotopes occurring in nature were created long ago in supernova explosions (in contrast to the continuously-created carbon-14 described above). The radiometric technique commonly used is to compare the abundance of the original isotope with that of its decay products, using the known decay rates. A favourite, which provides two clocks in one, is uranium-235 to lead-207 and uranium-238 to lead-206 (half-lives of 700 million and 4.5 billion years respectively). Accuracies of better than a few percent in age have been achieved. Others include potassium-40 to argon-40 (half-life 1.3 billion years), rubidium-87 to strontium-87 (half-life 50 billion years) and uranium-234 to thorium-230 (half-life 80,000 years). These half-lives (and many others) conveniently cover the entire history of the Earth.

Of course there can be complications. In the case of sedimentary rocks, are you measuring the time when the sediments were laid down, or the ages of the original specks of sand? Interpreting results for metamorphic rocks can be even more complicated. But there are several complementary dating methods. Many of them validate each other by criss-crossing over time. Fossils, as we shall see, follow time closely through their evolution. The outpourings of volcanoes provide thin but unmistakable time markers. The periodic reversals in the Earth's magnetic field keep score both on the seabed and on continents. The layering of rock sediments, punctuated by the odd outpouring of volcanic magma, define the temporal sequence just as well as a journal, even when inverted by crustal convulsions. Nevertheless, the radioactive clock, based on well-known physics, remains the ultimate foundation stone of the geological time scale.

Another kind of clock is provided by the mutation record in our genomes. An important finding in the 1960s, clearly predicted from evolutionary theory, is that, as species diverge from a common ancestor, the number of DNA changes (mutations) increases linearly with time. We can determine not only the sequence of branching events but also when they occurred. The accuracy of the method can be greatly increased by using many different genes or proteins, and the absolute calibration is provided by radioactive dating. Thus we may determine the timing of branching events even for organisms that have left no fossils.

A Brief History of Life on Earth

The earliest hint of possible life on Earth comes from 3.8 billion year old sedimentary rocks found in Greenland. These don't contain fossils, but they do contain pockets of pure carbon, which may conceivably be 'chemical fossils' of early organisms.

The oldest actual individual fossils, found in ancient sedimentary rocks in northwestern Australia and a few other locations, are 3.2–3.5 billion years old. They appear to be quite similar to today's single-celled cyanobacteria. In rocks about 3.4–3.5 billion years old, much larger fossils are found. These are the bulbous 'stromatolites', which can be several metres high. They are in fact vast colonies of ancient bacteria. They became much more abundant and conspicuous over the next billion years. Modern versions can be seen in shallow tropical waters today; they employ photosynthesis, suggesting, but not proving, that photosynthesis may have existed as early as 3.5 bya. (The terms 'billions of years ago' and 'millions of years ago' are abbreviated to bya and mya throughout this section.)

The atmospheric oxygen content is strongly related to life. Today it is about 21%, and this is maintained by photosynthesis. However, there are persuasive arguments that the atmosphere had only a small oxygen content until about 2 bya, when it markedly increased. This increase is attributed to an increase in photosynthetic activity, possibly due to the by then much more abundant stromatolites.

Aside from the bacteria and cyanobacteria, no other kinds of fossils have been found that existed before about 2.5 bya. That means that, as far as we know, microorganisms such as bacteria were the only forms of life on Earth for over two billion years – almost half the age of the Earth!

Somehow, about 1.4–2.1 bya, more complex organisms called eukaryotes appeared on the fossil scene. Eukaryotes contain a nucleus and other internal structures. Their development was very important, as they led to more complex cells and multicellular life. One might have expected this to have triggered a cascade of new and more complex organisms, but, aside from what may be some fossils of 1.3 billion year-old multicellular algae, no evidence has been found of complex organisms before about one bya. Why it took complex life forms so long to develop remains a mystery.

Sponges, mentioned above in Chap. 9, are important in the evolution story. They are invertebrates, and the earliest multicellular animals known, having first appeared over 600 mya. The genome of a sponge, containing some 18,000 genes, has recently been sequenced, and it says quite a lot about how individual cells first managed to live together. They had to find advantages in doing so, and they had to stick together, grow together, recognize their group as a unit and others as foreign. They also had mechanisms to suppress cells in their group that wanted to multiply at the expense of others (cancer was an early problem).

About 540 mya, the 'Cambrian explosion' occurred. Suddenly there was an abundance of fossils of shells and animals with skeletons, dramatic evidence of a rapid proliferation of a wide diversity of life. It was not just that animals developed hard parts that could easily be fossilized, as soft tissue can also be fossilized, and there are other kinds of 'fossils' such as burrows and footprints. There was truly a remarkable explosion in the diversity of creatures at this time, both soft-bodied and those with hard parts. Many factors may have coincided to make this happen. It was a period of great biological 'experimentation', with many extinctions as well as many successes.

The first vertebrates appeared about 500 mya. These were fish, and a wide variety eventually evolved. At least one of these was a precursor of terrestrial vertebrates. A good candidate is the lungfish, which still exists today. It can obtain oxygen either from water through gills or, when required if its pond dries up, from air through primitive lungs. The development of such features may have arisen as an adaptation to periods of drought.

About 380 mya vertebrate animals invaded the land for the first time. The first were, naturally enough, amphibians. Plants had already appeared on the land over 400 mya, and insects had followed close behind. All of these had to adapt to their new environment. Plants developed seeds (although flowering plants came on the scene much later, about 100 mya). The first insects were wingless, but flying insects similar to those familiar to us today developed over the next tens of millions of years. Soil containing decayed organic material became commonplace for the first time; large coal deposits have been found that date back to more than 300 mya. Colonization of the continents required the presence of ozone in the atmosphere to protect life from the deadly ultraviolet radiation of the Sun; the ozone layer increased along with the abundance of oxygen in the atmosphere.

Reptiles first appeared in the fossil record about 330 mya. This was at a time when many of today's continents were joined together in a large landmass, facilitating a wide radiation of species. The continents had hitherto been essentially barren compared to the biological activity in the seas and shallow waters. By 250 mya the populations of plants, insects, reptiles and other forms of life had multiplied to fill this new environment.

Then, about 250 mya, came the largest known mass extinction of life. It was a global event. It seems that no species was unaffected, and life on the Earth may have come close to total extinction. The specific cause(s) is still unknown. It should be noted that no period in history has been completely without extinctions; the process of evolution has always involved the replacement of some species with others over the course of millions of years. But the fossil record indicates that a number of exceptional extinction periods have taken place, some of them quite sudden on a geological timescale, even as short as a few thousand years. Each of these extinction episodes was followed by a period of rapid proliferation of new species.

The period following the mass extinction of 250 mya has been referred to as the age of reptiles. The most famous of course were the dinosaurs (from the Greek dino (terrible) and saur (lizard)). However the early part of that period was actually dominated by fairly large mammal-like reptiles, which were possibly warm-blooded and may have had hair. Their eventual extinction may have been due to dinosaurs, but some of their descendents, the true mammals, survived as small creatures to carry on the mammalian line.

The dinosaurs were around for over 170 million years (from about 240 mya until their extinction 66 mya), and were the dominant creatures for much of that time. They started small and gradually evolved, becoming larger and more diversified. They are by far the most famous of any extinct species, and the carnivorous Tyrannosaurus Rex and herbivorous 80-ton Brontosaurus are centrepieces in many museums around the world. The fact that they suddenly became extinct due to a massive impact from an extraterrestrial body only adds to the mystique of this extraordinary period in the history of life.

Reptiles known as the pterosaurs were also the first flying vertebrates, using flaps of stretched skin for flight (rather like bats today). They first appeared around 200 mya, and some of them eventually became huge, with wingspans of well over 10–15 m. Birds, employing a great new invention called feathers, followed about 50 million years later. All modern birds are descendents of dinosaurs.

The small mammals that somehow managed to exist through the epoch of the dinosaurs also somehow survived the extinction that eliminated the dinosaurs 66 mya. With the dinosaurs out of the way, the families of mammals expanded and diversified enormously. The primates date from that early time. There were and still are two major categories of mammals: the placental mammals (like ourselves), which have a long gestation period before birth, and marsupials, which are born at a much earlier stage of development.

Coming to what may be called recent times, DNA analysis has now more precisely determined the chronology. Divergence amongst the primates occurred in the following order: The macaque monkeys and the gibbons speciated from the great apes about 25–33 mya and 18–20 mya respectively. Amongst the great apes, the orang-utans speciated away 12–16 mya, the gorillas

6–8 mya, and the chimpanzees and bonobos speciated away from our own ancestral family 4.5–6 mya. Our ancestors living 3–4 mya were already bipedal; their footprints have been found in solidified volcanic ash dated at over 3.5 mya. It is thought that by then they had descended from the trees to the savannahs of Africa. Indirect evidence of tool use by our ancestors goes back to about 3.4 mya, and the oldest known stone tools are 2.5 million years old. Homo (our own branch of hominids, and Latin for 'human') also shows up in African fossils about 2.5 mya. Homo erectus had appeared in Africa by 1.8 mya. Fire was 'domesticated' about 1.5 mya. By a million years ago, Homo erectus had migrated out of Africa into Europe and Asia, and by 100,000 years ago, Neanderthals, with brains as large as ours, were living in Europe, and the recently discovered Denisovans were living in Siberia. Homo sapiens, our own species, originated about 200,000 years ago, and migrated out of Africa about 45–50 thousand years ago. The Neanderthals, who interbred with our direct ancestors to some extent according to genetic analysis but also probably suffered from the new competition, disappeared from the fossil record about 28,000 years ago.

The Ice Age of 20,000 years ago lowered sea levels, and would certainly have facilitated the migration of modern humans around the world, including the Americas. The beginning of civilization is considered to have started with the advent of agriculture some 10,000 years ago. The rest, as they say, is history.

The Evidence for Evolution in a Nutshell

The evidence for evolution is overwhelming. Like gravity it is an integral part of our lives, occurring all the time every day and all around us, without our even noticing. It is absolutely central in modern biology.

As many people think only of fossils when they think of the evidence for evolution, perhaps it is worthwhile here to briefly summarize the major lines of evidence for evolution.

The fossil evidence itself is understandably far stronger now than in Darwin's time, as many more fossils from a variety of epochs have been found. Evolution and the fragmentary fossil record make predictions. If two fossils are found from different epochs that seem to be part of an evolutionary sequence, there is a clear prediction that another fossil may eventually be found with intermediate morphological characteristics and an age between the original two. This has now happened on several occasions. The ages of the rocks within which the fossil resides can be accurately determined using radiometric dating, and, if the morphology of the newly found fossil is intermediate between the original two fossils, so invariably is the age. Gaps in the fossil record are being filled, in, and 'missing links' are being found.

However, the most dramatic, direct and compelling proof of evolution has come from a totally different direction than Darwin could possibly have imagined, a century after his Origin of Species was published. Every cell in our bodies (and in those of all other species) contains multiple records of our evolution. The breakthrough is due to molecular biology, and specifically genetics. As mentioned above, the copying of DNA, perfect as it is, can produce extremely rare errors (mutations). Mutations can also be produced on rare occasions by external factors such as natural radiation. Such a mutation can change a 'letter' in the DNA code forever, and the modification, small as it may be, will be passed down to all subsequent generations. For technical reasons, this is best detected using the Y chromosome which is passed down through males, and the mitochondrial DNA (mtDNA) which is passed down through females.

By sequencing an appropriate part of the DNA (or, as a proxy, the amino acid sequences in proteins) of a living organism today, we can look back over its entire heritage. And, in particular, when we compare different species in this way, we can see whether they share the same mutation in their DNA. If they do, then we know that they had a common ancestor. Thus, we can ultimately build up the tree of life for all species living today. This record should match the (much less complete) fossil record – and it does. Totally different methods give the same evolutionary history.

We know that the genome of the chimpanzee is 98.5% identical to ours, and that all human races are 99.9% identical. And our studies are not entirely restricted to the genomes of organisms living today. Most of the Neanderthal genome has been recovered, and is so close to ours that some scientists regard the Neanderthals and humans to be a single species. Most of the genome of the extinct woolly mammoth has also been recovered, and is similar to its close relative the African elephant within 0.6%. Even some properties of dinosaur genomes have been examined and compared to those of their descendents, the birds. Such studies hold much promise, and are still in their infancy.

The same methods of comparative DNA sequencing are being used, in finer detail, to trace our 'recent' human history (i.e. over the last tens of thousands of years). The Genographic Project of the National Geographic Society is sampling the DNA of hundreds of thousands of people all over the world, to trace their lineage and that of the groups they belong to, and ultimately to map out the great human migrations that emerged from Africa to all corners of the world.

Another line of evidence for evolution comes from embryology, the science of how organisms develop in the embryonic state. This in fact was Darwin's choice as the strongest evidence for evolution in his time. Even before the time of Darwin it was known that all vertebrate embryos start out looking very much like fish embryos. During development, major changes take place, and at birth the creatures are clearly members of totally different species, including reptiles, fish, birds and mammals. Human embryos have webbed hands and feet, and the fingers and toes eventually become distinct as the process of apoptosis (programmed cell death, described below) removes the material of the webs. In rare cases, some humans are actually born with webbed hands and feet.

Many other examples of differential embryonic development can be given. At one stage of embryonic development, human embryos are covered with hair, which is shed about a month before birth. The embryos of vertebrates start out with so-called branchial arches, which become gills in fish and major structures of the heads of mammals. The temporal order of these and other changes is reminiscent of the relative location of each species on the tree of life, as if the successive changes were evolutionary add-ons. This has all been studied in detail over the years, and amply demonstrates the unity of living systems, as it traces in real time the steps from almost indistinguishable embryos to fully developed distinct species.

The field of biogeography gives another powerful view of the effects of evolution. Darwin himself had an ideal opportunity to study it during his epic voyage on The Beagle. It is basically the study of the distribution of life over the surface of the planet. The variety and complexity are astonishing – perfect material for a biological detective.

Why do oceanic islands have such different populations from continents? Why are fossil seashells found in high mountains? Why are the desert plants in Australia, Africa and Asia so different from those in the Americas? Why are marsupials concentrated in Australia, while the earliest marsupial fossils are found in North America? Why are the flora of South America and South Africa so similar? Why are the fossils of some ancient trees found as widely dispersed as South America, Southern Africa, India, Australia and even Antarctica, when their seeds were too heavy to float? Why do regions with similar conditions (land, climate) have very different forms of life? The list of questions is huge.

Evolution has been able to answer these questions in detail (with some help from fundamental discoveries such as continental drift, which Darwin speculated about but which only became an established fact a hundred years later). A key example is provided by oceanic islands. These are islands that were formed by volcanic uplift from the ocean seafloor, and include the Galapagos and Hawaiian islands. Unlike continental islands (such as the British Isles, Madagascar and Tasmania), which contain the same species as the nearby mainland, the oceanic islands are devoid of the mammals, reptiles and fresh water fish that are commonplace on continents. But they are replete with species that are capable of long-distance dispersal (many plants, birds, and insects), and some that are found nowhere else. The rich field of biogeography has been and continues to be explored in all its complexity by evolutionary science.

Darwin began his Origin of Species with a chapter entitled "Variation under Domestication" (breeding by humans). This was meant to be a soft introduction to what he knew would be a far more contentious chapter, "Variation under Nature" (natural selection). Humans have been breeding plants and animals for only ten to fifteen thousand years – small compared with evolutionary timescales of millions of years – but the results have nonetheless been impressive. The domestication of grey wolves into dogs began about 10,000 years ago, and this has resulted in the greatest diversity in body size of any mammalian species: a factor of a hundred, from the 1 kg Chihuahua to the 100 kg Great Dane. The evidence for artificial selection is obvious, and the parallel to natural selection is clear. Similar success has been achieved for a wide range of plants and animals: horses, cows, pigs, flowers (think of the Dutch tulip mania), and crop plants of many species, especially cabbages.

Natural selection in the wild is normally thought of as being glacially slow by comparison, but we know of examples that have occurred within a human lifespan. One of the most famous involved the Galapagos finches, which endured a severe drought in 1977. These finches had to cope with larger and harder seeds for their food, and natural selection led to an increase of 10% in the average beak size within a generation – only larger billed finches survived. Many other examples of rapid natural selection have been documented.

If we really want to see extremely rapid evolution occurring today, right under our eyes, we only have to go to a biology laboratory or to a hospital. Replication times for bacterial species can be as short as 15 minutes, so we can observe evolution involving thousands of generations in real time. A single bacterium can replicate into a population of billions in just hours! A new field has developed to study these phenomena: experimental evolution. Some of the experiments have now been going on for decades. By comparing the evolved populations with the original parent populations which were kept in a frozen state and then revived, large changes have been observed. The changes induced include genuine speciation. In one case a bacterial strain was simply put into a beaker containing a nutrient broth, and observed over time. Because of an inevitable gradient in the properties of the broth (such as oxygen content) from top to bottom, within 10 days two new species of bacteria had evolved, one living in the top layers and the other at the bottom. Many organisms are suitable for such studies. Most fruit flies live for less than a month, so they are also ideal for population and evolutionary studies.

The most famous examples of rapid evolution in our environment involve diseases and drugs. Decades ago, antibiotics seemed to provide the miracle cure for bacterial diseases. However, as seen above, bacteria can evolve quickly. Mutations have made them resistant to many of our drugs, and new drugs are then needed to cope with the new strains (species). Many people may think that what has happened is that we are losing our resistance to the bacteria, but what is actually happening is that the bacteria themselves are evolving. The specific genetic changes in the bacteria have been identified by scientists – there is no question that this is true evolution. Other examples abound. One particularly well-known case is the adaptation of insects to DDT. Pathogens such as HIV evolving on a daily basis within their hosts provide another example, and dramatically highlight the need for an immune system as a rapid adaptive response. It is clear that we are engaged in an arms race with microorganisms – and all because of evolution, which really happens, with a vengeance.

The Timeline

It is instructive to look at the overall timescale, from the Big Bang 13.7 bya to the present. To put it all into easy perspective, using an updated version of the 'Sagan calendar', let us suppose that this entire period is compressed into just one year.

The Big Bang occurred at the stroke of midnight, at the very start of the new year. The first elements were formed less than a millionth of a second after midnight. Recombination and the formation of the microwave background took place 14 minutes later. The first stars and galaxies were formed within the first 2 weeks of January. Star and galaxy formation increased at a rapid pace, and the peak of this activity occurred in mid-March. Since then it has been declining gradually, although it continues into the present.

Our Sun and Earth were formed almost half a year later, at the beginning of September. The first tentative evidence of life on Earth appeared in mid to late September, and the first bacteria existed by late September. The large bacterial colonies called stromatolites were in evidence a couple of days later. Then nothing much happened until the first complex cells (the eukaryotes) and the first multicellular algae came into existence in late November.

After another relatively uneventful period, the first invertebrates appeared on December 13th. The Cambrian explosion, when a large number of new organisms appeared, occurred on December 17th. Then things really started to move. The first vertebrates appeared on the 18th, plants, insects and vertebrates had invaded the land by the 21st, and the first reptiles appeared on the 23rd. A major period of extinction occurred 2 days later, but life recovered and many new species came onto the scene: dinosaurs, mammals, birds and a great number of others. December 30th was the terrible day when a large asteroid collided with the Earth, killing off the dinosaurs and many other species. Mammals were amongst the species that survived.

Coming to the very recent past, our ancestors the hominids appeared 2 hours before midnight on New Years Eve. Homo erectus migrated out of Africa about 40 minutes to midnight, and the Neanderthals lived in Europe just 4 minutes to midnight. Modern humans, our direct ancestors, made their famous migration out of Africa 2 minutes before midnight. Human agriculture started about 20 seconds to midnight, recorded history about 10 seconds to midnight, and radio technology was developed two tenths of a second before midnight.

Our Earth entered the cosmic picture rather late in the game, and most of action of life on Earth only happened in December, particularly in the second half of that month. We humans appeared on the scene in just the last few minutes, and what we consider to be our advanced technology only appeared in the last fraction of a second.

13. Comings and Goings

No account of evolution would be complete without a consideration of the comings and goings of individuals on the stage of life.

In the grand sweep of things, as we look at the overall evolution of life, we think of the histories of entire species and families of species. But what about the countless individuals that comprise those species? An overview of life must certainly include the start and end points for individuals (especially ourselves): conception and death.

Why Sex?

There are essentially two ways of reproducing: sexually and asexually. The vast majority (over 99%) of plants and animals reproduce sexually. Why? What are the advantages of sexual reproduction? A definitive answer still eludes us, but there are many hypotheses.

Asexual reproduction is simple: it just involves a cell splitting in two, resulting in two new cells identical to each other and to the original. One hundred percent of the genes of the mother cell are passed on to the daughter cells. That's about as good as you can get.

By contrast, there is a very significant cost for sexual reproduction. Only half of a mother's genes are passed down to future generations; the other half comes from the father. So each parent has passed on only 50% of its full complement of genes. This is referred to as the twofold cost of sex (or, if you like, the absurd cost of males). But it's even worse. Much time and energy is spent locating and attracting mates, avoiding predation in the process, and often caring for the young. In fact, these activities seem to totally dominate the lives of many animals around us. The disadvantages of sexual reproduction seem to be overwhelming.

P. Shaver, *Cosmic Heritage*, DOI 10.1007/978-3-642-20261-2_13,
© Springer-Verlag Berlin Heidelberg 2011

Yet sexual reproduction is almost completely dominant amongst plants and animals. What advantages are there that can possibly outweigh the huge costs?

Darwin, as usual, was there early. His view, expressed in 1889, was that 'hybrid vigour' in the offspring was sufficient to justify sexual reproduction. In the same year, August Weismann had the important insight that sexual reproduction produces the genetic variation required for natural selection and evolution. Even in the face of the many competing hypotheses proposed over the past century, this is still widely considered to be (one of) the most important advantages of sexual reproduction.

There are certainly a number of reasons why the mixing produced by sexual reproduction can be advantageous. Re-shuffling the genetic deck increases randomness and the genetic diversity of new generations. It creates unique individuals. Producing many unique offspring in an unpredictable environment can make it more likely that at least some of them will survive. Any benefits from the new combinations of genes can ultimately be shared by all members of the species. It is true that sex can also break down combinations of genes that work – but these are combinations that work at the moment, and they may not work in the future. It is important to have a large library of combinations that can be called upon at any time and in unforeseeable circumstances. Sexual reproduction provides a mechanism for genetic repair: damage to one strand of DNA can be repaired by reference to the other, complementary, strand. A population may also more readily be able to purge itself of harmful mutations, through natural selection.

An important issue that has been much discussed over the last several decades has to do with our never-ending battle with parasites. It is called the "Red Queen" hypothesis, and has been championed by William Hamilton amongst others. Parasites are always trying to break down our defences, and we are constantly modifying our defences to evade the parasites. The situation is reminiscent of the Red Queen in Lewis Caroll's *Through the Looking-Glass*, who has to run just to stay in one place. We and the parasites are in an eternal arms race, neither one of us getting significantly ahead, so we effectively go nowhere. Of course, if the parasites could ever crack our defences, we would be stopped dead in our tracks, and the race would be over.

On the face of it, the parasites have a huge advantage over us. They reproduce rapidly, and so can mutate much faster. The AIDS virus evolves constantly, while a human generation is 20–30 years. The only way to keep up with a constantly-changing parasite is to possess a huge in-house library of genetic combinations that can be called upon at random by the immune system whenever needed. Just a fraction of our 21,000 genes, through different combinations, can produce millions of variations in the immune cells. Sex is necessary to replenish and re-shuffle this library from time to time, and the result is that each of us has a unique set of combinations. Our offspring will again have a new and totally unique combination of parasitic resistance alleles. It is important that we keep all of the alleles in our library, even those that do not seen to be particularly useful at the time, as we need to be able to shuffle an enormous number, and we never know when a particular allele will be called into service. This vast array, and its combinatorial potential, is what keeps us in the game.

Parasites are constantly trying to break into cells. To do so, they have to evade the immense flexibility of the immune system in producing antibodies, and they have to crack the passwords of the cells (which are known by the immune system so that it does not attack the host's own cells). The parasites are constantly inventing new keys, and the body is constantly changing the locks. It is an eternal stalemate – as long as our defences can keep up. A sexual species has a vast library from which to produce ever-changing antibodies and locks. Sexual reproduction therefore seems to be essential for our survival, given the ever-present menace of parasites.

So there do seem to be a number of possible advantages of sexual reproduction, including greater genetic variation, the spread of advantageous traits and beneficial mutations, the elimination of deleterious mutations, the creation of new allelic combinations, DNA repair and the fight against parasites. While biologists may argue whether these are beneficial at the level of the group, the individual or the single gene, overall they do seem likely to be beneficial. But do they really overcome the twofold cost of sex?

It does seem reasonable to suggest that asexual reproduction may be better in a constant environment, while sexual reproduction may be better in a diverse, changing environment ('environment' here includes both the inanimate and the biological). In the extreme case of a static, homogeneous environment with an unlimited food supply, asexual reproduction can in principle proceed without any impediment. One cell replicates itself in two identical cells, those two each replicate in the same way giving rise to a total of four, and so on *ad infinitum*. In no time at all there are billions of identical cells from the original one. The only problem is that all of these cells have the same genetic composition, and so can cope with just the one single environment; if there's any unfavourable change in the environment, the entire population can rapidly become extinct. So an extreme asexually reproducing population can grow very rapidly in a favourable, static environment, but can be entirely wiped out by a single unfavourable change. In reality, of course, it is not quite so simple. Mutations occur rapidly because of the short generation times, and there are other ways in which genetic material can be modified and even transferred. So asexually reproducing populations do not stand still – far from it, as we have seen. And bacteria have been successful on this planet for 3.5 billion years. However, the huge diversity of unique individuals resulting from sexual reproduction might make it even more likely that at least some individuals, and therefore the population, will survive environmental challenges. Together with the other advantages mentioned above, the case for sexual reproduction may perhaps start to look promising.

What about solid evidence? It is surprisingly hard to come by, considering how preponderant sex is. But an increasing number of observations and studies now seem to support the main hypotheses. Eukaryotes, which are mostly sexual, have evolved much more than bacteria, which are asexual. The genes coding for immune system proteins evolve considerably faster than genes coding for other proteins, as expected from the Red Queen hypothesis. Studies involving species that can reproduce either sexually or asexually have been informative. Some such species reproduce asexually in the summer, but sexually in the harsher winter conditions. In one recent study it was found that the cloned versions of certain snails in New Zealand suffered major losses when they were infected with parasites, whereas the sexual snail populations remained stable. In another recent study it was found that a species of rotifer shifts to sexual reproduction when the environment becomes changeable. The experimental work in this field will certainly increase, as this is the only way to finally be sure about the reasons why sexual reproduction is so predominant.

Do We Have to Die?

Is death of the individual a necessary and pre-programmed part of evolution? Or is it just the inevitable breakdown of a complex machine?

These are two extreme views on death. According to the first, we may be programmed to die. Even if we could somehow survive the diseases and accidents that plague us throughout life, there may be an instruction written in our genomes (a 'death gene') that ultimately and unavoidably causes death by a certain age. According to the second, it is just the random mutations and the gradual 'wear and tear' in the cells that occurs over the years that cause ageing and death. In either case, in an ideal world devoid of diseases, accidents and predators, a fundamental question is whether natural death ('dying of old age') is inevitable.

A mechanism for programmed cell death does indeed exist. The process is known as apoptosis. Paradoxically, it is an important part of the formation and maintenance of the human body, and is well known to biologists. Apoptosis occurs in response to signals which may come from outside or inside the cell. It kills cancerous cells, infected cells, damaged cells, and early cells that are not part of the final body plan (as mentioned above, we all start with embryonic webbed hands and feet, and, *in utero*, apoptosis eliminates the material between our fingers and toes).

The elaborate network of signals tell the cell to commit suicide. They activate specific enzymes to cut up the contents of the cell, including its DNA. The remnants are then consumed by scavenger cells. The process of apoptosis protects neighbouring cells from damage. Apoptosis occurs in many eukaryotes, indicating that it arose early in their evolution.

While it is an important process for life, apoptosis can also have negative consequences. It has been implicated in some degenerative diseases such as Parkinson's and Alzheimer's, and some cancers. However, it seems unlikely that apoptosis, by itself, causes ageing and death in general. It is an agent, not a cause.

There is another process, at the genetic level, that certainly does set limits on cell life. It is called replicative senescence. In the early 1960s Leonard Hayflick discovered that there is an upper limit to the number of times a normal cell can divide (fifty or so, in his studies). Each time a strand of DNA is copied, the ends are not fully reproduced, and the DNA becomes shorter. The effect is mitigated by the presence of lengths of non-coding repetitive ('junk') sequences called telomeres located at the ends of the DNA; the fraying shortens the telomeres without damaging the coding DNA (the telomeres serve the same purpose as the plastic ends of shoelaces). However, when the telomeres are eventually depleted, the fraying affects the coding DNA and the cell dies. This is referred to as the 'Hayflick limit'. On the face of it, this is a ticking clock that ultimately kills cells and could conceivably be a major cause of ageing and death.

But there is an enzyme called telomerase that can reverse this process and restore the telomeres on each division. This has been proven by inserting into normal cells a gene that activates telomerase: the lifespans of the cells are increased. Because of telomerase the cells can actually become immortal, at least with respect to replicative senescence. Stem cells are effectively immortal for just this reason. Unfortunately, so are cancer cells, in which the telomerase seems to be reactivated. Telomerase 'takes the brakes off' and facilitates cancer – but if we can turn telomerase off for these specific cells we may be able to stop cancer growth. Replicative senescence may be regarded as a mechanism to prevent the onset of cancer.

Whether replicative senescence plays a significant causal role in ageing has been a matter of some debate over the years. The evidence is accumulating, as discussed below.

Several general theories of ageing have been proposed by biologists over the years. One rather obvious one was proposed by August Weismann in 1889. From the point of view of evolution, our deaths would appear to be essential. The old must make way for the young who can reproduce. Reproduction not only produces new members of our species, but it also makes possible the genetic variation which is essential for evolution. A population that cannot evolve to adapt to the continually changing environment will die out.

However, evolution is not teleological. Evolution works from past causes, not towards future goals. So while death may be convenient for evolution, there is no overall forward-looking 'plan' of which it is a part. Furthermore, this theory would imply that the individual's self-interest is sacrificed for the good of the species as a whole. This goes against some modern views in which the interests of the 'selfish gene' are put first.

Another straightforward theory put forward in the early 1900s was the 'rate of living hypothesis', in which it was proposed that a high metabolic rate could lead to short maximum life spans. It has subsequently been found that this theory is often in contradiction with observed differences in lifespan between species, or even amongst members of individual species.

In 1952 Peter Medawar proposed the influential 'mutation accumulation theory', according to which natural selection works only up to the reproductive age, and after that it does not. Ageing is simply caused by neglect over the later years. There would be no pressure from natural selection to keep the body going after the reproductive age, especially after disease, predators and accidents have taken their toll anyway. There may also be random harmful mutations that have an effect only late in life. These late-acting mutations could accumulate in populations over evolutionary time. Alternatively, even moderate senescence can preferentially kill off the older members of a population through predation and disease in a competitive environment. The old, being less healthy and fit than the young, preferentially fall victim to predators and disease.

George Williams continued this general line of thought with the theory of antagonistic pleiotropy, which he proposed in 1957. In pleiotropy, a single gene may influence more than one trait; it is 'antagonistic' when one of the traits is positive and another is negative. An example suggested by Williams was a gene that could promote calcium deposition in bones (which is favourable for early development), but which could also promote calcium deposition in the arteries (which is unfavourable in later life).

Accordingly, Medawar's scenario would be due to genes that give a selective advantage in youth, but are harmful in later life.

In 1977 Tom Kirkwood put forward a rather different view: the disposable soma theory. The body must allocate its limited energy resources to various tasks: metabolism, reproduction and repair. According to the theory, the reproductive cells take precedence over the body repair function, and the body (soma) cells gradually deteriorate with time. Ageing is then the result of preferentially investing resources in reproduction rather than in maintenance of the body.

All of these theories encounter various experimental problems, but overall there seems to be a broad consensus that ageing is due in large part to random events related to molecular disorders (often mutations), and is not genetically programmed. This is consistent with the common view that the body simply declines with age. Skin tone and elasticity degenerate, with visible consequences given the never-ending pull of gravity, and there are many other obvious results of 'the machine' being run for a long time. Natural selection promotes health up to the reproductive years, but after that we are 'on our own'. As Hayflick put it, "Longevity ... is a measure of how far we are able to 'coast' on the excess physiological capacity or redundancy remaining after reproductive maturity."

As mentioned at the beginning of this section, we are concerned here with the ultimate, fundamental causes of ageing and death – those that remain after we exclude disease, accidents and predation (humans are now largely free of death by predation, while it remains one of the major causes of death for other species). The former determine the maximum life span. By contrast, the average life span (or life expectancy) is considerably less, dominated as it is by disease and accidents.

The longest confirmed life span known for a human is 122 years (Jeanne Calment, a French woman who lived from 1875 to 1997). As far as we can tell, the maximum life span of humans has remained essentially constant at about 110–120 years over recorded history, whereas the average life span has increased considerably, especially over the last 100 years, due to improvements in health and disease control. The average life span is estimated to have been 18 years during the bronze age,

20–30 years in Greek and Roman times and in medieval Europe, and 30–40 years just 100 years ago, compared to 80–83 years in several countries today.

According to the World Health Organization, 58 million people died in 2005. The ten leading causes of death in the U.S. in that year were the following: heart disease (27%), cancer (23%), stroke (6%), chronic lower respiratory diseases (5%), accidents (5%), diabetes (3%), Alzheimer's disease (3%), flu and pneumonia (3%), kidney disease (2%), and septicaemia (1%). One hundred and thirteen causes of death were listed in the National Vital Statistics Report, but 'natural causes' or 'old age' did not appear in the list (they never do). What if all diseases and accidents were eliminated? The only remaining cause of death would then presumably be 'natural causes' (i.e. systemic/genetic breakdown, finally causing death).

Leonard Hayflick has pointed out that "There is ample evidence that our longevity is influenced strongly by our genetic heritage... 48% of 90-year olds and 53% of centenarians had both parents who lived to 70 or older. These percentages are significantly higher than those for people who died at younger ages." Or, as Elizabeth Blackburn once commented, "If you can get past what kills most of us – get through what I call the hail of bullets – then your genes can do you some good." Genetic influences are also clear from the fact that identical twins tend to die within three years of each other, compared to 6 years for fraternal twins.

The role of genes in determining life span is also obvious from a comparison between the (approximate) maximum life spans of different species. The maximum life span is generally determined either from the mean life span of the most long-lived 10% of a species, or from the life span of the oldest known member of that species. Here is a list of typical *maximum* life spans: 15–60 days (fruit flies), 30 days (mosquitoes), 50 days (houseflies), and, in years: 2 (moles), 4 (mice), 6 (Earthworms), 15 (beavers), 25 (dolphins), 29 (dogs), 36 (cats), 49 (goldfish), 62 (horses), 76 (chimpanzees), 86 (elephants), 90 (sponges), 122 (humans), 190 (tortoises), 200 (whales), 410 (molluscs), 4800 (bristlecone pine trees).

But genes are not the whole story. It has been found that genetically identical laboratory animals raised in identical environments have significantly different life spans. This is consistent with random mutations and molecular disorders playing a role in longevity. An increasingly diverse range of biological factors involving the cells both individually and collectively are being found to influence ageing, as might be expected in any complex machine. Certainly no single fundamental factor underlying all the known diseases and causes of ageing and death is known. Some of the many putative influences on ageing are mentioned below.

Oxidative stress has long been the most popular explanation for ageing at the molecular level. It is caused by the system's inability to cope with and detoxify the reactive oxygen products such as oxygen free radicals, and to repair the resultant damage. Normally, antioxidant enzymes maintain a balance, but this balance can become destabilized, and oxidative stress has been implicated in many diseases. However, the relative importance of oxidative stress has been called into question in some recent studies.

The mitochondria, the powerhouses in the cells, are particularly susceptible to damage over the years. They contain their own DNA, which is totally independent of the DNA in the cell's nucleus, but the mitochondrial DNA is more fragile, has no efficient repair mechanism, and is very sensitive to mutations. Furthermore, a side-product of the mitochondrial energy production is the release of the damaging oxygen free radicals mentioned above. All in all, the result is the gradual deterioration of the energy production by the mitochondria, particularly in the nervous system and muscle cells. Ultimately cell death can result, through the process of apoptosis. Some computations of the deterioration of the mitochondria with age are consistent with the present estimate of the maximum human lifespan.

Inflammation is one of the body's responses to potential hazards ranging from invasive pathogens to internal irregularities. It provides protection, but can also cause serious problems, so it is normally closely regulated by the body. As infectious disease has always been a major killer, it is no wonder that we have such an aggressive immune response system. But it is increasingly being realized that long-term side effects can accumulate and contribute to ageing and death. Inflammation may play a role in many of the illnesses of old age. This may then be another form of antagonistic pleiotropy: what saves us up to the reproductive years may ultimately contribute to our deaths.

The two leading causes of death mentioned above are complex. 'Heart disease' is an umbrella term for a plethora of complications affecting the heart and vascular system. One of the many involves a build-up of plaques on the walls of arteries that contain cholesterol and calcium amongst other substances. This is one of the most obvious examples of accumulating damage over time. Cancer goes to the very core of life: it involves alterations of the genome itself. In that case it may seem somewhat surprising that the ultimate causes of cancer are largely environmental, not hereditary. Mutations to the genome caused by the environment can disrupt the normal cell mechanisms and lead to runaway growth and invasion of other tissues.

Which takes us back to the telomeres. As mentioned above, the shortening of the telomeres in our DNA is certainly a possible contributor to ageing and death, and perhaps the closest thing to a real 'ticking time bomb'. The enzyme telomerase can mitigate this effect. But in the extreme case of cancer cells, telomerase can completely 'take the brakes off', allowing uncontrolled growth.

It has been reported that telomerase, the enzyme that restores telomere length, is most active in youth; in later years the telomeres become progressively shorter. A typical cell of an infant has of the order of 5–15,000 repeating telomeric base pairs; this is reduced by about 50 base pairs per year on average. When the telomeres disappear the cell can no longer replicate properly, and it dies. There is some evidence that the telomere rate of shortening in other species varies and affects their life spans; some long-lived sea birds seem to be particularly immune. There is increasing (but often tentative) evidence that in humans many factors, including several mental and physical disorders, oxidative stress and psychological stress, contribute to the shortening of the telomeres. Recent twin studies suggest that telomere length is dependent on environmental factors rather than heritable factors, and that relative telomere length can actually predict which twin is likely to die first. It now seems likely that telomere length is indeed a significant factor in the ageing process.

A fascinating new study indicates a possible link between some of these proposed causes of ageing. It seems that a protein which is activated to stop cell division when the telomeres become too short also has the side-effect of repressing critical genes in the mitochondria, thereby reducing the function and number of these vital energy-producing bodies. Even worse, this may enhance the damaging reactive oxygen products, and a vicious cycle may result. Thus, telomere dysfunction may be related to mitochondrial degeneration, oxidative stress and decreased metabolism. Not a pretty picture.

These are just some of the many proposed contributors to ageing and death. It is clearly a complicated matter. But research in the area continues to advance rapidly as we learn more about the detailed processes of life.

Suspended animation isn't quite the same thing as death, but it is relevant and instructive. Many living organisms can survive over long time periods at temperatures below 0°C. Some species of bacteria have reportedly been revived after being frozen in ice for thousands of years, and some extremophiles can survive in a dehydrated state at close to absolute zero for up to a decade. We all know that many plants can survive in freezing temperatures for months, and some vertebrates can do so for weeks. How does this work?

The field of cryobiology studies the effects of low temperatures on living organisms, and covers refrigeration, hypothermia, hibernation, the preservation of human eggs and sperm, the freezing of cells to very low temperatures and much more. Human embryos have been frozen for periods up to many years and then thawed and implanted, with no increase in abnormalities. In a living organism, as the temperature is reduced, there is simply less and less energy to drive molecular motions, and cellular activity decreases. At the limit of absolute zero, there is no energy and no activity; the material substances of the cell simply continue to exist in a state of limbo, like so many inert grains of sand. Years later, if the temperature increases so does the molecular activity, and the motions of life return.

In practice, however, it is not quite so simple, and considerable effort has gone into making the process safe. A major problem is the water in our bodies and cells. At 0°C the water freezes and expands. This can cause massive damage to other components of the cell, which would otherwise do just fine at lower temperatures.

There can also be phase changes in membrane lipids, and other complications. The challenge is to somehow keep the cell safe as the temperature drops through a critical range (from above the freezing point of water to significantly below). Beyond that range, even lower temperatures can actually provide protection for the cell for long periods of time. Careful control over the freezing process is required to get through the critical range. An alternative approach is to avoid the freezing and thawing of water altogether with the addition of certain chemicals. At very low temperatures, well below the critical range, a 'vitrified' system can result – a 'solid liquid', or 'glass'. Molecular motions virtually come to a stop. This provides a vivid demonstration of the fact that a living system is comprised of a collection of molecules that can do just fine over long periods of time at very low temperatures; it is the activity, which requires energy and therefore temperature, which makes them 'alive'. Simply add heat, the molecules move, and hey, presto – there's life again! It is the same activity that leads to degeneration over time, and ultimately death.

Are there really maximum life spans? A 'mortality plateau' has been found for some species, such as fruit flies. After a certain age the death rate stops increasing: it levels off and then stays roughly constant. This is a complication for some of the theories of ageing mentioned above. But an important implication may be that there is no fixed upper limit to longevity.

OK then – what about immortality? There are some potentially immortal species. Bacteria may be considered to be immortal, but only as a colony. These single-celled creatures reproduce through cell division, producing identical daughter cells. As this process goes on through generations, the overall colony could be considered to be potentially immortal. Some bacteria can reduce themselves to 'endospores' when times get bad: an endospore just contains the DNA and a few cell constituents, encased in a protective coat. As in the case of low-temperature preservation, there is no metabolic activity. The endospore is very tough, resistant to desiccation, high temperature and extreme freezing, and ultraviolet radiation. Endospores can exist for thousands, possibly even millions of years, and then, when conditions finally improve, reactivate themselves into the original bacterial state. This is impressive, but it involves a suspension of life, not a prolongation. We're more concerned with ongoing life.

Hydras are simple fresh-water animals whose cells continually divide; they appear to age slowly, if at all. A mature jellyfish known as *Turritopsis nutricula* can revert to a stage of immaturity (the polyp stage) through the cell conversion process of transdifferentiation; this cycle can conceivably repeat itself indefinitely, resulting in biological immortality. And it has been speculated that the ancient bristlecone pines mentioned above may be potentially immortal. At the cell level, stem cells and cancer cells are said to be potentially immortal. On the other hand, there are famous cases of apparent 'programmed death', such as the Pacific salmon, which dies in the stream where it was born within days of laying its eggs, having previously spent several adult years out in the open ocean.

What can be done for humans? Medical and biological scientists are striving to increase both the average life span and the maximum life span in a variety of ways, ranging from controlling disease to making fundamental changes through molecular engineering. A conservative objective is to maximize the healthy years of life (the 'health span') within the overall constraints of the maximum life span; this is sometimes referred to as 'squaring the curve'. Eliminating all causes of death presently appearing on death certificates would add as much as 14–15 years to life expectancy – a huge achievement, but not as much as the typical 25-year increase over the last 100 years, and still not reaching what is presently the maximum life span. However, many seek to extend the maximum life span through different avenues, perhaps achieving rejuvenation in some areas. And some even dream of the ultimate goal, immortality.

It has often been said that we don't want to live for a long time anyway. We become poor in health, life is awkward and painful, our friends have died and our grandchildren have grown up and don't need us anymore. It can be a very lonely existence. (Of course, 'squaring the curve' can help: healthy and active lives while we live, then sudden death.) Many people have the same view of extremely long life – just more and more dismal frailty – or of being 'unfrozen' and waking up in the totally alien environment of the future. Again lonely and unwanted. But supposing we *all* lived, at a fixed stage of development (corresponding, say, to 30 years of age) for thousands of years together. It would seem normal to all of us. Society would adjust to fit the reality. We would have serial careers. Population would have to be controlled, but we could still have children from time to time through our long lives. Death would come largely from accidents. People might take care of themselves a bit better if the stakes were thousands of years rather just another five or ten. It is not a totally crazy scenario, and need not be at all unpleasant, although it may be considered extremely optimistic. Leonard Hayflick has expressed the opposite view, and thinks that immortality, if it could happen, would be unimaginably unpleasant. An interesting topic for a walk along the beach, but at present the prospects seem impossibly remote, and in reality we can only focus on marginally increasing our average lifespans.

A final comment on the subject of death. It has been suggested that we may be the only creatures on this planet that can ponder their origins and contemplate their own deaths. According to this view, death by whatever means just happens to other creatures without their anticipation that their lives will end, that they will cease to exist. Of course they do all they can to avoid predation, and have many built-in or learned responses to avoid anything that might injure them or cause sickness. But they may not have the conscious awareness that they are thereby trying to avoid death. Perhaps our ancestors were in the same position just a few million years ago, before conscious awareness developed fully.

Contemplation of one's own ultimate death (or that of a loved one) is certainly not pleasant. In this respect we may actually be worse off than other creatures ("ignorance is bliss"). Perhaps sometime in the future (a hundred, a thousand, a million years?) we may have solved the problem of death in some way. But in the meantime we're stuck with an awareness of inevitable death. If so, as life on Earth goes back 3–4 billion years, we're presently in the (unfortunate) position of living in a very special period – a tiny fraction of the history of life – when we can contemplate our own deaths but can't do much about it.

But we can contemplate much more than our deaths. The same mental abilities have enabled us to contemplate the distant universe and the complexities of life around us. They have inspired us to work together towards great goals. They have given us a remarkable human spirit that enables us to rise above adversity. Our mental abilities are undoubtedly what make us exceptional on this planet.

But did they evolve, like everything else? That is the subject of the next chapter.

14. Did Cognition Evolve?

Evolution of the Brain

Over the past millennia many believed that the very cores of our beings – our cognition, our minds, our consciousness – reside in our hearts. Now of course we know that our brains have this distinction.

In 1838, two decades before he published his *Origins of Species*, and when he was cautiously probing his ideas, Darwin wrote in his notes that "the problem of the mind cannot be solved by attacking the citadel itself. – The mind is a function of body". Only much later, in 1872, did he grapple with the problem of the mind in a book entitled "The Expression of the Emotions of Man and Animals". There he argued that both humans and other animals express the same state of mind by the same movements. Darwin had no idea of the mechanism for natural selection related to physical characteristics, let alone the mind. But we now have detailed knowledge of the unity of life for all species – the underlying genetic basis that provides for natural selection, the genetic code itself, and the mechanism that powers life. Can we do the same for mental capacities? This is the topic of the present chapter.

How far down the tree of life do we find brains? By definition one-celled organisms cannot have brains – unless the entire organism is a neuron! Actually, some unicellular eukaryotes do have remarkable capabilities for such apparently simple organisms, and one of them, the *Paramecium*, is sometimes referred to as the 'swimming neuron'. Some multicellular eukaryotes, such as the sponges, do not have a nervous system, while others, such as the starfish and the jellyfish, have a decentralized nervous system with no brain.

P. Shaver, *Cosmic Heritage*, DOI 10.1007/978-3-642-20261-2_14,
© Springer-Verlag Berlin Heidelberg 2011

Most invertebrates have brains. These include insects, crust-aceans, octopuses and squids. As invertebrates have been around for well over half a billion years, it has been assumed that they were the first to have brains. One well-studied invertebrate is *C. elegans*, a tiny worm with just 302 neurons. Another is the well-known fruit fly *Drosophila*. Much has been learned about some of these brains, and that knowledge is relevant to human brains.

Most relevant to human brains, however, are the vertebrates, all of which have brains. Simple anatomical examination reveals a striking and obvious pattern of evolution. Mammalian brains in particular share a common architecture with the human brain. As Elman et al. (1996) wrote, "There is no evidence that humans have evolved new neuronal types and/or new forms of neural circuitry, new layers, new neurochemicals, or new areas that have no homologue in the primate brain." In other words, the ingredients are identical. The common fundamental components are the forebrain, the midbrain, and the hindbrain, which is con-nected to the spinal cord. The most basic functions involve input of sensory information and output in the form of motor functions. In mammals there is a well-established correlation between brain size and body size, with some scatter related to mammalian type. The human brain lies further above the correlation line than others. The correlation is obviously due in part to the needs of the body, but it may also be related to behavioural aspects, such as the size and complexity of the social group.

The most obvious part of the mammalian brain is the cortex (also common to other mammals), which covers other parts of the brain and has a well-known characteristic convoluted and crum-pled appearance. If the human cortex is smoothed and spread out, it is seen to be a sheet 2–5 mm thick covering an area equivalent to that of about four sheets of typing paper. It has to be crumpled in order to fit inside the skull. It has six main layers (which originated in early mammalian times), but vast networks of connections span the entire thickness. The outer layers are grey due to the many neurons or cells (the 'grey matter'), but its interior is white due to the mass of axons that carry signals between the neurons. The brain has two distinct hemispheres, which are joined together by a large bundle of nerves. Buried under the cortex are several other well-known regions of the brain, which will be outlined in the next chapter.

The neurons of the human brain have been known for over one hundred years. They are also present in the brains of other vertebrates, and invertebrates such as insects and worms. When neurons were first discovered, it was realized that they are extremely unlike any other cell in the body. While most of the organelles are confined to the cell body, as in any other cell, there is also an astonishing 'tree' of complex structures radiating away from the central body of the cell. These are dendrites that receive signals from other neurons. In a typical neuron there is also a single axon extending from the central cell body, which transmits signals to the other neurons. An axon ends in a number of branches, the input interfaces of synapses which transmit the input information to the receptor. (Some axons – those extending down the spinal cord – can be up to a metre long, even though they are only a thousandth of a millimetre thick; scaling up, that corresponds to a 4-m diameter subway tube extending from London to Cairo – quite a commute!). The synapse is a gap across which chemicals called neurotransmitters pass information from the transmitting neuron to the receiving neuron. It is now known that there can be well over a thousand proteins associated with a given mammalian synapse.

It came as a surprise that some single-celled organisms with no neurons at all contain surprising amounts of synapse proteins. Perhaps the origins of the brain go back a long way indeed – to proteins that unicellular organisms use. The number of synapse proteins seemed to increase with evolutionary time, twice as many in invertebrate synapses, and twice again as many in vertebrates. The synapses themselves also changed. This may be an indication of one of the ways the brain evolved, and the time scale over which it evolved.

At the higher levels of animal cognition, we normally think of the great apes (Homimidae), which include humans, chimpanzees, bonobos, gorillas, and orang-utans. Most people would certainly not think of birds as being amongst the more intelligent animals. Indeed, the phrase 'bird brain' in common usage is derogatory, implying stupidity and minimal brainpower. This couldn't be further from the truth. Some birds are surprisingly intelligent. The historical reason that 'bird brain' became a derogatory term was, at least in part, that anatomical examination showed that the brains of birds contain a very small cortex. The cerebral cortex in mammals is generally considered to be where the highest brain functions reside. However, it is now known that avian brains have a fundamentally different structure. Birds have a structure called the hyperstriatum in their forebrains that performs the same functions as the cortex in mammals. Unlike the folded and convoluted cortex of mammals, the hyperstriatum in birds is rather amorphous in shape. The neurons are scattered throughout this structure, rather than being located in specific layers as in the mammalian cortex. Birds fit quite well into the brain-to-body mass correlation mentioned above for mammals.

Thus we have learned quite a lot about brains over the years – both our own brains and those of other creatures. Brains have been around for over half a billion years, and all share the same basic elements. Ours is not by any means unique, although it is certainly unusually powerful. The evolution of the brain is just as continuous as any other aspect of evolution. In the next sections we will look at animal behaviour, and what it tells us about our mental capacities.

Innate Behaviour

At one extreme of types of behaviour are simple reflex (or 'knee-jerk') reactions, immediate responses to external stimuli. At the other extreme are learned behaviours and thought processes, most obvious in humans. Between these two extremes lies an enormous spectrum. The extent to which humans and other animals overlap in these behaviours has been hotly debated for years.

We can easily understand basic reflex actions. They can be based on simple physical or chemical stimuli, and are seen in very basic organisms. They can easily be simulated by robots or computers. We share this primeval capability with virtually all life forms. It may be interesting to ask where the boundary is between simple reflexes and complicated instincts, but in reality there is probably a continuum between the two.

An instinctive behaviour has normally been considered to be one that is automatic, irresistible, of a fixed pattern (unmodifiable), unlearned, and common to all members of a species.

Instincts in animals can be extremely complicated indeed: feeding habits, migrations, mating behaviour, care of the young, singing and other means of communication... the list is endless, and fills entire textbooks on animal behaviour. A few examples are given here to illustrate some of the incredibly complex behaviours generally considered to be instincts.

A digger wasp prepares for the future life of its offspring by digging a burrow, paralyzing a caterpillar, placing it in the burrow, and planting an egg on it so the baby wasp has food while growing. The wasp has to be able to find its way back to the burrow with the caterpillar, and will check the burrow before dragging the caterpillar into it. If the paralysed caterpillar is moved (by some mischievous researcher) while the wasp is checking the burrow, the wasp will move the caterpillar back and then check the burrow again as thoroughly as before. The unnecessary second checking of the burrow is taken as evidence that the entire process is purely instinctive (robotic), with no thought involved. But it is certainly remarkable that all of this is done by the wasp alone, with no training from its parents, who it never meets – instinctive actions can be amazing and non-trivial.

Many animals have elaborate courtship rituals, and the common fruit fly is one of them. The male goes through its standard song-and-dance routine, which includes tapping the female and vibrating its wings to produce a characteristic sound, before proceeding with copulation. It has been shown that a female fruit fly, provided by researchers with a single male-type gene, will then also go through precisely the same courtship ritual with a female virgin, although of course in this case copulation does not follow. The presence of one particular gene has caused a female to behave exactly like a male, and in a highly complex and specific manner. Something quite similar has been found in the case of mice. Functional neuronal networks of both male and female sexual behaviours exist in the brain of each sex, and can be switched on or off by gender-specific sensory modulators. Entire behavioural patterns, involving highly specific actions, are stored, ready to go, in the material brain.

In some social insects such as ants and bees, there are 'genetic switches' that become effective at some stage in the life cycle to determine an individual's role in life. In the case of honeybees, researchers have recently succeeded in actually controlling which larvae will develop into queens and which into workers, essentially 'at the flick of a switch'. In one case 72% of the larvae turned into queens, and in another other case 77% turned into workers! In principle they could make *all* the larvae into queens, while normally there is only one queen in a hive of tens of thousands of bees – the actual fraction they achieve just depends on technical issues. In nature, when worker bees decide to make a new queen, they choose several larvae and virtually immerse them in a nutritious substance called royal jelly. After sequencing the honeybee genome the researchers deduced that DNA methylation (described in Chap. 10) may be involved. Once they knew this, they were able to mimic the effects of royal jelly by making changes directly to the methylated genome. The genomes of queen larvae and worker larvae are identical, so the methylation (which results in the same genome being expressed in different ways) can make the difference. And what a difference! The behaviours are completely different, and the queen outlives the workers by a factor of over 20. Complete behaviour patterns, including 'knowledge' of the (yet-to-be encountered) environment, result from the genome and modifications to its expression.

With direct evidence like this, there can be no doubt that instinctive behaviour is genetically based. This by itself is a truly remarkable fact. But how can the expressed genome possibly code for behaviour of such incredible complexity, requiring such an exquisite knowledge of the environment? It is already quite remarkable for the expressed genome to code for the entire (but fixed) body plan and morphology of an animal, something that is unvarying over its lifetime. But it would seem to require vastly more to code for dynamic interactions with the world at large, apparently requiring a knowledge of that world in all its varied dimensions and perspectives.

Consider a young deer that finds herself being chased by a wolf. She runs towards a thicket, jumps over a log and dashes through some hanging leaves. How does she know immediately that the log is solid and the leaves are not? Is such detailed information innate? How can it be? What is the relationship between the genome and behaviour (especially innate behaviour)? The genome by itself would presumably be inadequate, but genes can set up quite complex and specific neural networks in the brain, and it is these that must provide the elements of innate behaviour. Nevertheless, animals armed only with blunt innate abilities surely don't have enough to cope completely with the intricacies of the real world. The young deer needs time after birth under the protection of its mother to acquire an adequate knowledge of its environment. Gerald Edelman has said that the brain encounters the real world with a clash. Experience has to be gained quickly. The first few days are critical.

Do Humans Have Instincts?

This may now seem like a strange question, but for much of the past century many have regarded the human brain as a 'blank slate' (at least with regard to personality traits and abilities), devoid of any innate tendencies or abilities, and with no possibility of instinctual behaviour. In this extreme view the brain of a young child is totally empty, to be moulded entirely by human culture and learning. Certainly it is true that learning and culture have played major roles in our lives. However, we do indeed have instincts, in spite of their presence being masked or smothered in many cases by learning and culture.

Examples abound. The reality of instinct in human physical behaviour is obvious. In human babies, sucking, clasping, crying, sitting up, standing, walking and climbing can all be classified as instincts. The determination of a one-year old to get up and walk on two feet is impressive, and that baby will never revert to previous behaviour. Hunger, gagging, diarrhoea, and excreting are all parts of the normal and emergency elements of metabolism. Taste is important in our selection of food; sweet things tend to be nutritious and bitter things tend to be toxic. The biochemistry of taste is well known. Sexual urges and behaviours are essential for human procreation. Sleeping is essential for life (although amazingly we still do not know exactly why we spend a third of our lifetimes on this (in)activity). Fright and flight responses (or aggressive responses) are important for survival. We harbour fears of many kinds, including primeval phobias. Research over the years has revealed a wide variety of innate characteristics, ranging from purely physical behaviours to personality traits and abilities as outlined below, which are involved in many aspects of our lives.

Studies of identical twins have been extremely important in establishing the reality of inherited traits and abilities. Their advantage is that behaviours and preferences can be explored that are based on broad swathes of the genome, giving results that are far superior to studies that concentrate on the effects of just one gene.

Identical twins are formed when a single egg is fertilized and then divides into two separate embryos. Both embryos have identical DNA (although environmental conditions both inside the uterus and throughout later life can produce minor variations in behaviour). About 0.4% of human births produce identical twins. Fraternal (non-identical) twins form when two independently fertilized eggs occur in the uterus at the same time. Unlike identical twins they share only 50% of their genes, and are no more similar than any other siblings. The comparison of identical and fraternal twins ('classical twin studies') can obviously reveal a great deal about inherited factors. A wide variety of other tests using twins can be construed. One of the most powerful is the comparison of identical twins reared apart. As they are identical genetically but have lived lives in very different environments, they are ideal for studies of nature versus nurture. Another approach is to study individual pairs of twins over time. Many other such studies are possible. With modern computers, sophisticated statistical techniques and large databases around the world that total more than 600,000 twins, the results are very robust.

And they are absolutely clear. Virtually every characteristic studied has a strong genetic component. Genetic factors determine 40–50% of the variation in all of the major personality traits (agreeableness, conscientiousness, extroversion, neuroticism and openness) indicies of psychopathology, and many others. They determine over 80% of the variation in IQ (for adults). A recent large study in The Netherlands has shown that heritability is responsible for 50–90% of the variation in a range of talents: music, art, writing, language, chess, mathematics, sports, memory and knowledge. Even the sense of fairness is more similar between identical twins than between fraternal twins. Twin studies have also been critical in showing the high level of heritability in several categories of disease (both mental and physical).

A novel experiment in Europe has recently combined twin studies with functional magnetic resonance imaging (fMRI) technology. Identical twins and non-twin brothers were given a memory task which was interrupted by an unrelated 'distraction' task, while their brains were being scanned by the imager. In this way a complex cognitive process was followed in real time for each of the subjects, and the results were later compared. It was found that, while there was considerable variation in how the brain of each subject dealt with the problem, the individual pairs of identical twins strongly tended to use the same neural networks as each other. The estimated heritability ranged from 60% to 90% for the three components of the test. Clearly, with innovative experiments such as this, twin studies have a big future in understanding genetic processes.

Where does instinct end and thought begin? As in all these things, it seems to be a matter of degree – a continuum. Many of the complex behaviours we attribute to instinct may be considered to involve thought – or even to require thought. Where do we draw the line? Or is that arbitrary? Do many behaviours involve elements of both instinct and thought?

Most scientists now agree that both nature and nurture are generally involved – not just one or the other. And that there is an interplay between the two. If that is the case for humans, presumably it is also the case for other animals.

Complex Behaviour

There are many cases known in which animal behaviour appears to go far beyond mere robotic actions. Many of these show adaptability, variation, and versatility – totally different from some of the defining characteristics of instinct (automatic, repetitive, unvarying and unmodifiable). Can some of these versatile behaviours be considered to be evidence of advanced cognition? Cognition can involve mental activities such as planning, judging, choosing, problem solving, reasoning, recalling memories, reflection, imagining, having goals, having a comprehension or model of the world, and, in the case of consciousness, a concept of self. Can any non-human animals do any of these things? The difficulty, of course, given that non-human animals can't speak, is that any evidence is circumstantial only, and so we cannot reach absolute conclusions, either for or against advanced animal cognition.

Countless possible examples have been given. The use of tools by sea otters, which use stones as hammers and anvils, and by assassin bugs which employ camouflage and bait; coordinated hunting by lions, hyenas, wolves and dolphins; communication, warnings and deception by many different species. There is even evidence that many mammals have REM (rapid eye movement) during sleep; in the case of humans, REM sleep is an indicator of dreaming. There are many examples of animal behaviour that appear to go well beyond instinct.

Learning is another aspect. We are all aware of the well-known learning capabilities of most if not all mammals. Perhaps not so well known is the fact that insects can also be trained. Remarkable cognitive abilities that go well beyond mere robotic instincts are found well down the chain in the animal kingdom. Some examples are given here.

Honeybees

Insects like honeybees, with small and supposedly simple brains (only 950,000 neurons in the case of honeybees), have a significant learning capacity and cognitive ability, and their behaviour is not solely (or even mostly) instinctive. There is considerable plasticity in the insect brain; it is activity-dependent, not hard-wired. Experience of the world after birth is very important. For honeybees, some of the clues come from smell, polarized light, touch and even the shapes of flowers; they are capable of learning what is most important to them. They can be taught Pavlovian responses, and can even be trained to take a specific direction when faced with maze-like 'crossroads'.

An outstanding example is the famous 'waggle dance' of honeybees. When a worker bee returns to the hive having just discovered an exceptional source of food, she communicates this information to her co-workers inside the pitch-dark hive. This she does by performing a so-called waggle dance. She walks rapidly in a straight line while waggling her abdomen back and forth, and then circles back and repeats the performance again and again. The direction of the straight waggling run is correlated with the direction to the new source of food relative to the direction of the Sun. The length of the straight waggling run is related to the distance to the food. And the vigour (degree of agitation) of the waggle dance is related to the desirability of the food supply. This is certainly a form of symbolic communication.

The waggle dance ('wanzltanz') was discovered by Karl von Frisch, an Austrian working in Munich during the second world war. When the bombing of Munich became too severe, he and his researchers moved the experiment into the countryside, with the unanticipated result that the accuracies achieved in confirming the waggle dance became much greater because of the longer distances available for the honeybees to explore. His meticulously researched studies initially met with a storm of disbelief and criticism when they were announced, but the waggle dance of honeybees is now widely considered to be the most sophisticated example of non-primate communication known.

Honeybees have demonstrated some amazing abilities for communication and learning. This photo shows the queen (the larger bee marked with a white spot), which lives for years, amongst some of her tens of thousands of workers, which live for just a couple of months (Photo by Ryszard Maleszka, Australian National University).

Octopuses

It makes its way across the open soft-sediment sea bottom in a most bizarre way, in a lumbering gait with stiffened legs ('stilt walking') while carrying coconut shell halves. It's a Veined Octopus. Julian Finn, from Australia's Museum Victoria, and colleagues have observed this behaviour repeatedly over the past decade, off the coast of Indonesia.

The open sea bottom is a dangerous environment for these fully-exposed creatures. An opportunity to provide some protection for themselves came up as they discovered coconut shell halves that had been discarded by humans. Several times they have been observed carrying the shell halves (having first used jets of water to clear them of mud) over distances of up to 20 m, and then using them for shelter. When an octopus only has one half of a coconut shell, it will just pull the shell over itself. When it has both halves, it plants one open-side up, clambers in, and then pulls the other (open-side down) over itself.

Clearly, carrying these shell halves out in the open in this awkward way involves a risk, as the octopus is far more exposed to predation. But it gives a *deferred benefit*, ultimately providing an effective hiding place and shelter. This is tool use, which is impressive enough in itself. But taking an immediate risk to use a tool for a deferred benefit is certainly on a higher scale of the cognitive ladder.

Invertebrates have been known to show behavioural flexibility indicative of cognitive ability, but the sophisticated behaviour described above takes it to a whole new level. It is interesting to note that tool use was once considered unique to humans. But the more we study and observe, the more we realize that it is not at all unique to us, and that in fact cognition itself may well be a continuum extending from insects to us.

Dolphin Strand Feeding

First the birds come flying in, landing on the beach grasses along the sandy shoreline. They wait in anticipation.

Then suddenly a concentrated tidal wave explodes out of the calm sea, comprised of water, dolphins and myriad small fish. Pandemonium reigns. The fish desperately flip and flop into the air and back onto the sandy shore. The dolphins are on their sides, thrashing to and fro, catching the fish between their sharp teeth in a feeding frenzy. As the meal dissipates, the dolphins wriggle themselves back into the sea to regroup. The birds now swoop down to clean up what remains of the fish.

It is astonishing. How did the birds know where and when to gather? How did the dolphins decide as a group to corral the fish into a tight ball and herd them onto the sand. Who gave the order? How did they ever develop such a strategy? Presumably this strategy was 'discovered' by the dolphins over generations, and the young have learned from the old. But it can certainly never be called naïve.

Humans also developed such strategies (actually somewhat simpler strategies) over the millennia, corralling and stampeding wildebeast or buffalo to plunge to their deaths over a cliff. But humans likely developed the technique much as dolphins did, rather than planning it theoretically from first principles in an office.

Bower Birds

Of course we all know about peacocks, and the extraordinary tails of the males that serve to attract females while at the same time putting the male, with its heavy and colourful plumage, in mortal danger from predators. This is just one of a vast number of strategies by which males vie for mating rights.

An even more amazing but less well known example of male strategies in the mating game is provided by the bower birds, all seventeen species of which are found in New Guinea and Australia. In this case the male bower bird constructs a complex structure (the bower) whose sole purpose is to attract a female partner. The female meticulously examines one bower after another, in order to select its mate. The bower is not used as a nest; it is abandoned after courtship has been successfully completed, and the real nest is then built somewhere else. The bower is probably the only structure in nature that is constructed purely for sexual selection, and begs the question of whether some sense of aesthetics may be involved.

The basic structures built by a given species are of the same grand design, but they are by no means identical. They range from simple avenues comprised of twigs, sticks and grasses, to raised avenues, to avenues crossed at right angles, to 'maypoles', to rather complete and impressive huts. The maypoles can be as high as 3 m, the huts as wide as 2 m, and the thickness of the avenue walls greater than 20–30 cm. The overall weight of a bower can be greater than 60 times the weight of the bird itself. The bowers are impressive structures.

The basic structure of a bower is just the beginning. It is then decorated, with enormous care. The decorations can be elegantly simple, as in the case of the dark blue and purple flowers, feathers and berries used by the satin bowerbirds. At the other extreme they can be extravagantly elaborate, including berries, leaves, bones, beetle wings, snail shells, flowers – and even human debris such as bottle tops and other shiny objects that may be found discarded in the area. In some cases the ornaments are sorted and placed according to colour or some other criterion. Again, significant variations are observed, suggesting that the patterns are not purely innate. And – could overall variations that are sometimes seen over decades even suggest fads? In any case, there are certainly significant innovations, which suggest something beyond purely innate behaviour.

These bowerbird mating activities are not quite as innocent as they may seem. In fact they are in deadly earnest. In one geographical area just 15% of the males achieved over half of the matings. For the genes of the males, this is a matter of do or die. So it is not surprising that the males pay almost as much attention to the bowers of their competitors as they do to their own. Given a chance, they will destroy the competitors' bowers. At the same time they are highly protective of their own.

Does bower building, with its innovation and flexibility, suggest a certain level of cognitive ability? And, given that it is entirely useless except as a display for attracting a mate, might it even tell us something about the nature and origin of aesthetics? For those who would argue that it is merely an innate behaviour resulting from a specific type of natural selection, the same might be said of various aspects of the human sense of aesthetics.

Beavers

Perhaps the ultimate example of adaptable behaviour comes from the industrious engineering activities of the beaver. The beaver has such an extended kindom to control that it has to be prepared to be flexible in what it does. Its home is partly above and partly below the water of the lake it created by building a large dam in a stream. It maintains burrows to the shore and channels that lead into the forest that it harvests for its building materials and food. The channels can be deep enough to float tree branches to storage piles. The beaver is constantly on the lookout for breaks in the dam, which it must repair immediately. It can't stand the sound of trickling water. On the other hand it will actually let some water escape during winter, which leaves a convenient air gap between the water level and the ice above, so that the beavers can freely swim about while the wolves prowl around on the ice above. Beavers use the most appropriate materials available to them for building and repair. It is hard to imagine that they can do all this without some anticipation of the results of their endeavours, and without some concept of their far-flung empires.

Corvids

The corvids, which include jays, crows, ravens, magpies and others, are considered to be the most intelligent of the birds. Their brains contain an unusual number of neurons. Magpies are one of the few species to have passed the mirror test for self-awareness (see below), and crows have demonstrated remarkable tool-making abilities. But first we will consider some amazing talents exhibited by the western scrub-jays.

The research described here is largely due to Nicky Clayton and colleagues at Cambridge. She first observed the behaviour of the scrub-jays during lunch breaks on the lawns of the University of California, Davis. But rather than just casually noting their presence as most of us would, she started to take great interest in exactly what they were doing. The more she saw, the more intrigued she was, and this ultimately led to a meticulous and extensive research programme with very significant implications.

The basic behaviour was rather straightforward: if a jay had obtained more food than it needed at the moment, it would cache (store and hide) the excess for future feeding. It would 'simply' have to remember where it had cached the food. But then it gets interesting. Several caching strategies seem to be unique to the corvids, and they are the only cachers that can remember the locations of caches made by others. The jay is aware of the presence of other jays ('conspecifics'), which could conceivably steal the food if they saw where it was cached. If there were no conspecifics around, it would just cache the food and leave it until it was later needed. However, if there were conspecifics nearby which could observe the caching, the jay would re-cache the food at the earliest opportunity. In some cases the food would be re-cached five or six times. Each time the cacher would have to be able to recover the new cache with the same precision as previously.

An interesting twist is that only the jays that had themselves previously been thieves would re-cache, and only when they had been observed caching. Jays with no thieving experience did not re-cache. The implication is that jays with thieving experience infer the possibility that conspecifics may also steal, given the opportunity ("It takes a thief to know one"). Another subtlety is that the jays remember which conspecific observed them caching, and modify their re-caching behaviour in accordance with that information. They are much more likely to re-cache if the conspecific observer is a dominant bird in the group. And they are unlikely to re-cache if the conspecific observer is their own partner, with which it would normally share food in any case. Re-caching is much more likely if the original caching took place in well-lit conditions rather than in dark conditions.

Auditory information is also used. Jays are careful to be very quiet when caching under conditions in which the conspecific could hear but not see the caching event. By comparison, when the jay was alone, or when it was being observed visually by a conspecific, no effort is made to reduce the noise of the caching event.

These studies are of considerable importance in the field of animal cognition. They add to the evidence that some animals use knowledge of events of the past to plan future actions. They may also have implications for the question of whether any nonhuman animals might have some of the elements of a 'theory of mind' – the notion that others may have cognitive processes similar to one's own ('mental attribution'), making it possible to understand the viewpoint of the other individual (empathy in some cases, Machiavellian deviousness in others). In humans this ability develops around the third year of life, but it is obviously difficult to prove that it may exist in nonhuman animals, which can't communicate with us.

Another example of corvid cognition comes from a different direction – the construction and use of tools by New Caledonian crows. They have developed two types of tool to get at hidden insects and larvae. Both involve natural hooks and barbs that can be adapted from native plants to fashion tools that can be scraped in crannies where the insects and larvae are found. The prey can then be consumed out in the open. A sceptic may argue that this whole procedure is at least partially innate. But a wonderful example that is certainly not innate comes from experiments in the lab. A small bucket (with a handle) containing food was placed at the bottom of a clear plastic cylinder. The crows were given straight lengths of wire. In each case the crow first tried to extract the food bucket using the straight piece of wire held in its beak, but to no avail. Then, amazingly, the crow managed to bend the wire to make a hook, and proceeded to lift the food bucket out of the cylinder using the innovated hook. There can be little question that this behaviour involves both insight and planning. However there are two curiosities here: only the female crows succeeded in this (and they did so in 90% of the cases), but the males never succeeded. The other curiosity is that the females always tried first with the straight wire before finding the solution by bending it in various ways.

Yet another corvid has demonstrated its wiles. Ravens were perched on horizontal branches, and pieces of meat were tied to long strings suspended from the same branches. It was useless to fly and attempt to grab the securely-tied morsels, and the ravens couldn't just pull up the long string. But they found that if they pulled the string up in stages with their beaks, each time stepping on the string so that it did not fall down again, they could eventually gain access to their objective. This could certainly not arise from any innate behaviour; the solution had to be invented.

One could go on and on about the cognitive abilities of other animals. Chimpanzees are our closest living relatives. The others in our line have all become extinct. That's a great pity, as we would have learned a great deal about the evolution of our own cognitive abilities by studying them. The chimpanzees parted ways with us some 5–6 million years ago, and since then they and we have evolved separately. Nevertheless, studying them has been extremely informative, and countless volumes have been written about them. Suffice it to say that impressive cognitive abilities have been amply demonstrated. From these and the other less-known examples given above, the point is clear:

Continuity of Behaviour

So far in this chapter we've looked at what is known about the evolution of the brain itself. We've considered innate behaviour, showing that even very complex animal instincts result from the instructions of the physical genome, and that instincts also abound in humans. We then explored what is known about animal cognition, showing that at least some other animals seem to have significant cognitive capabilities. (In the next chapter we will see that some non-human animals even show evidence of self-awareness.) While Darwin was restricted to "The Expression of the Emotions of Man and Animals" in arguing for a continuity between the two, we now, through decades of patient observation and experimentation, have far, far more evidence to show that there is indeed such a continuity. The continuity is seen both genetically (through the evolution of the brain) and in a wide variety of behaviours. Just as the genetic code, the cell and the Krebs cycle are common to all life there also appears to be a continuity in behaviour and cognition throughout. The categories of behaviour appear to overlap and blend, and it is hard to define absolute boundaries. The blurred continuity could be one of complexity: from common reflexes to complex instincts to learned and adaptive behaviours and to thoughts. Although there may be a continuum, it is not necessarily a straight line, and at some point our mental capabilities may have increased far faster than those of other animals.

The Great Brain Explosion

It is amazing to think how recently our exceptional cognitive abilities developed. The wheel was only invented 6,000 years ago. Agriculture only started 10,000 years ago. What was life like for our ancestors who lived a few million years ago? How did they compare with some of the non-human animals we've just been considering? Just five million years ago the achievements of our ancestors were probably little more than those of many other animals. Actually they were probably dwarfed by the vast achievements of the social insects, by the engineering ingenuity of beavers, and by the cleverness of the corvids. Our early ancestors may have lived and hunted in groups, but so did several other animals. What clue would there have been at that time that they would become so exceptional? When did humans begin to 'over-achieve'? How did humans develop their extraordinary mental capacities that now seem to be so far beyond those of other animals? Ultimately it may come down to the size and complexity of our brains.

Our ancestors living 2–3 million years ago still had small brains (300–400 cm^3, similar in size to those of chimpanzees today). But then things started to happen. Brains started to increase significantly in size about two million years ago, and the rate of increase seems to have accelerated about 0.5–1 million years ago. Overall, the brains of our forefathers tripled in size in a comparatively short period of just a couple of million years. Our brains are now significantly 'over-sized': they are about 3.5 times larger than they 'should be' for a mammal of our size. And some brain areas are more over-sized than others. It is perhaps of particular significance that our cerebral cortex grew much more than the rest of the brain, and the prefrontal cortex, the seat of our highest cognitive powers, grew even faster than that. Relative to our cortex, our prefrontal cortex is twice that in chimpanzees (in absolute terms it is seven times bigger). Some other areas of the cortex are similarly over-sized, such as Wernicke's area, which is crucial for language. So although the ingredients of the human brain are the same as those of other mammals, the human brain significantly increased in overall size relative to the others, and specific areas of the human brain (especially those related to 'higher functions') increased even faster. And a huge multiplier effect in the number and complexity of brain connections could have resulted from the increased brain volume. There are suggestions that certain regions became more specialized in function, and that the prefrontal cortex became more connected to the rest of the brain. It has been a remarkable development. Could one say that, when quantitative changes become large enough, the effect becomes almost equivalent to a qualitative step (similar to co-called emergent properties)?

The reason for this remarkable evolutionary jump is unknown. Considering the costs of such a large brain in terms of birth, rearing and energy requirements, the evolutionary advantage must have been huge. In addition to our overall mental capacities, the development of language took place over the same period, and may have been crucial in multiplying our capabilities. The combination of these extraordinary developments may explain why our mental capacities go well beyond those of other animals.

The human brain contains some hundred billion neurons (nerve cells), each of which is connected to about a thousand others. It has been said that the number of potential connections in the human brain is roughly equivalent to the number of elementary particles in the entire universe. The human brain consumes about 20% of our total energy intake (50% in the case of infants!). The large size of our brains likely explains why human childbirth is so difficult, and why human babies are 'born prematurely' in that they are so helpless at birth and have to be cared for over such a long period of time.

The large human brain needs time to grow, and it does indeed appear that cognitive abilities develop gradually in young humans. In studying infant humans, scientists face the same problem as when searching for cognition, self-awareness and consciousness in other animals: they can't talk. From birth babies obviously have the instincts required for survival, but the degree of self-awareness or consciousness they have is unknown, and is likely small at the beginning. Memory itself also develops over time. At ages less than a year, children can only remember events from the previous month or so. By the age of two they can remember events for up to six months or a year. The quality of the memories also seems to increase with age. It is clear from many studies that the development of language plays a major role in the development of memory. In the case of long-term memory there is a remarkable phenomenon referred to as childhood amnesia. As adults we cannot remember events prior to the age of about three or four. There is some sort of barrier. It has nothing to do with the potential reach of memory, as elderly people can easily remember events from their teens. It is probably more related to developing the ability to establish long-term memories at those ages than to subsequently losing memory. It may also be somehow related to the full development of self-awareness and consciousness, which may emerge at those ages. The long period of continued development of the human brain after birth may be a major advantage.

In any case, the human brain has certainly become exceptional. And there have been other extraordinary (and probably related) evolutionary developments:

The Great Evolution Explosion?

It has been suggested on the basis of recent genetic studies that the rate of human evolution has been rapidly accelerating over the past thousands of years, and may now be as much as a hundred times faster than it was a few million years ago.

As noted above, we and the chimpanzees parted ways some five to six million years ago. Our ancestors were making tools two million years ago. Some sort of primitive communication evolved. Gradually, innovations and new skills were being developed. Some theories hold that a 'Great Leap Forward' took place in human development about 50,000 years ago, when modern humans migrated out of Africa, while others see a more gradual development. In either case, there is no question that our species has experienced extraordinary change over the past few million years. Is it possible that the rate of change has actually been rapidly accelerating over just the past several thousand years?

It is well known from plant and animal breeding that significant evolutionary change is indeed possible in much less than 10,000 years. Just look at the variety of dogs (first domesticated from grey wolves about 10,000 years ago) and cabbages (which have been bred into many varieties of plants). Substantial changes are possible in just a few tens of generations. Even a change in a single 'letter' of the genome can have a big effect. We now have direct evidence from modern statistical genomic studies that some human mutations are only several thousand years old. A famous example is a mutation that makes it possible for adults to digest milk (they are 'lactose tolerant'). This mutation originally arose about 8,000 years ago. Today 85–100% of northern Europeans are lactose tolerant, while only 5% of Chinese are. Adult lactose tolerance became important for the mobility of a society, as well as providing a food of enhanced nutritional value. It probably had a huge effect on history. Another important mutation was one that conferred significant immunity to malaria in Africa. It is thought to have originated about 2,500 years ago.

How was this rapid evolution possible? One factor was the huge increase in population. Before the exodus from Africa 50,000 years ago there were just two or three hundred thousand humans. A few thousand years ago there were about 60 million. Far more mutations were introduced, providing far more genetic variation. Beneficial mutations which used to occur every 100,000 years were showing up every few hundred years. And beneficial mutations can spread very fast, even in a large population. Another way of acquiring new genes was through interbreeding with the Neanderthals, which is thought to have occurred to some extent. Rapid evolution is possible without new mutations or new genes; in a changing environment the shuffling of the existing genome will result in the most appropriate alleles and gene combinations becoming dominant. Several of the new alleles found in recent genetic surveys are related to the brain. Some of them have effects on mood and emotion, and others contribute to aspects of brain development.

What other changes would have stimulated such rapid evolution? Major changes were probably due to the activities of humans themselves. They migrated all over the globe, and had to be able to adapt to their new environments. For those who migrated to northern latitudes, a lighter skin colour was essential to capture sufficient vitamin D from the reduced sunlight. For the Tibetans and for the Amerindians of the altiplano, physiological modifications were essential for survival at high altitudes (recent studies indicate that many of the genetic adaptations to high altitude amongst the Tibetans may have arisen in just the last 3,000 years – further evidence of fast recent evolution). But the change with the greatest impact may have been innovation. Increased innovation – a mental activity – would have caused pressure to compete, and ultimately to evolve in cognitive abilities. A larger population is more capable of producing more innovations, and, with language and trade, these spread very fast. There is increased pressure on societies to adapt as fast as possible, in order to survive the competition. Certainly the most important and far-reaching innovation was the advent of agriculture, about 10,000 years ago. It provided much more food than foraging, and made large populations possible. It gave stability, and led to civilization and culture. Science itself would have benefited from stable societies and increased connectivity, as it is ultimately a social enterprise.

But, in spite of the huge evolution and the many beneficial changes that have taken place, is it possible that we still live with some elements of 'stone-age brains'? It seems unlikely that our brains have been *totally* transformed. Children still harbour deep-seated primeval fears of the shapes of spiders and snakes, as demonstrated by endless psychology tests, while they don't have the same gut fear of the shapes of cars, which kill far more children than spiders do! Our preference for sweet-tasting foods over bitter ones probably goes back a long way; it would be a natural preference in raw nature, as bitter things tend to be poisonous. We obviously have a continuing capacity for fighting and aggression. What selection pressure would have made us less competitive, less aggressive over the years? Living together in an agricultural society may have had some influence, but it was obviously not enough – and increased competitive innovation would have worked in the opposite direction. While we are increasingly working together as a 'global village' to solve problems and differences with increased communication and civilized institutions such as the United Nations, the very innovations that have propelled us forward over the millennia have quite likely increased competition, so the same evolutionary pressures that applied millions of years ago probably still apply today. War itself would be a driver in evolution (just think of the famous expression "war is the mother of invention"), and it is reasonable to think that the same competitive traits still exist in the gene pool. The resulting aggression may be increasingly tempered by civilization, but can still lead to genocide in extreme cases. The twentieth century has probably been the bloodiest in history.

On the brighter side, a positive example of possible recent cognitive evolution, according to a hypothesis put forward by Gregory Cochran and Henry Harpending (2009), may come from the Ashkenazi Jews, who have lived in Europe for over a thousand years. Because of various restrictions placed on them, they had to concentrate on trade and finance, both requiring more intelligence than most of the other occupations at that time. As they were a relatively closed society, there would have been genetic selection resulting from competitive pressures in these occupations, and it is conceivable that this may have caused an increase in their overall intelligence relative to that of other groups (while the average difference may not be great, a small shift in the overall curve could make a big difference at the high end of the scale). Supporting evidence comes from their DNA. They have some unusual genetic diseases (such as Tay-Sachs disease) that can cause increased neural connections in the brain. Cochran and Harpending argue that the short timescale is feasible, while other disagree. But in any case many think that cognitive abilities may be increasingly driving our evolution.

Not all scientists believe that rapid evolution continues to take place. On one hand, it has been argued that modern life now protects us from the environmental pressures for evolution. On the other hand, we may well be on the brink of the fastest rate of evolution the planet has ever seen, due to bioengineering. The debate continues.

Looking back over this chapter, the range of behavioural abilities in living systems is impressive – from innate behaviour to complex behaviour to subtle behaviours that demonstrate innovative cognitive abilities, and even to compelling evidence for a 'theory of mind' in non-human animals. We have undoubtedly only scratched the surface, and have much to learn about the nature of cognition in many species. And there is one outstanding phenomenon that we have not considered yet – consciousness.

15. What About Consciousness?

What Is Consciousness?

We all have a profound and private sense of self and identity, which persists throughout life. Our past and future seem to blend together into a 'stream of consciousness'. Consciousness is the totality of our sensations, emotions, memories, values, tastes, curiosities, thoughts, opinions, beliefs, ideas, decisions, and of course self-awareness. The most amazing thing about it is its oneness: It is everything in one word. It is what we feel is 'in control' of our lives. The 'conscious self' has a worldview, and makes plans and initiates actions accordingly. The list of what we may consider to be aspects of consciousness is very, very long.

Our consciousness is a completely private thing. No one else can ever experience it. Nor can we experience anyone else's consciousness. Indeed, how can you even say for sure that the colour 'red' that you see is exactly the same as the colour 'red' that someone else sees – or even that that person is conscious? In this sense we all live in our own worlds, although of course we don't normally think of it that way.

The nature of consciousness has been and still remains a mystery. Is consciousness really created by the firing of neurons in the brain? How can the physical world (the brain) possibly produce such a comprehensive realm of intensely personal and private experiences? This topic has long been a subject of philosophical discussion. For most of the past century it was a 'forbidden topic' in science, as it seemed inaccessible to the scientific method, but in recent years, particularly with rapid developments in neuroscience, it has come to be considered a subject potentially amenable to scientific investigation.

P. Shaver, *Cosmic Heritage*, DOI 10.1007/978-3-642-20261-2_15,
© Springer-Verlag Berlin Heidelberg 2011

Did Consciousness Evolve?

It is easy to see why a primitive form of consciousness, in the simplest sense of just 'being aware', might have evolved. Animals can cope with the world a lot better if they are conscious rather than unconscious. They can avoid predation better, they can hunt better, they can mix in their societies better. Whether the particular aspect of 'consciousness' involves the senses, decision-making, or the subtleties of society, it is clearly better to have it than not.

At the most fundamental level, an early form of organism could perhaps be regarded as a colony of still smaller and more basic organisms that found synergies in working together. Over time, it would obviously have been beneficial if this colony came to regard itself as a unit. Simply moving about in an environment of obstacles would seem to imply an own-body concept. Therefore, the evolution of some primitive form of self-awareness would seem to be a natural result of cooperation in complex organisms.

A much more sophisticated level of self-awareness and ultimately consciousness would have become a huge advantage in social animals, and may well have evolved in line with the complexity of the society. The development of a 'theory of mind', in which one realizes that others also have consciousness, and can imagine what they may be thinking ("I know that you know that I know..."), can be an obvious advantage in the competitive world of a complex society. Indeed, the brain sizes of primates are correlated with the sizes of the groups they live in.

Are other animals self-aware? It's difficult to know, if they can't talk to us. Tests have been proposed, and one in particular has become especially well known: the mirror test. Darwin made some observations of animal expressions using a mirror, and in 1970 George Gallup developed a quite rigorous test to see whether animals can recognize themselves in a mirror – presumably an indication of self-awareness. It involves placing a coloured mark (which is odourless and cannot be felt) on or near the animal's forehead when it is asleep (or anaesthetized), and noting whether they try to touch or remove it when they later see themselves in a mirror. As a control, invisible marks can also be applied, to make sure that it is only the visible mark that attracts the animal's attention.

Chimpanzees were the first to be given the mirror test. It was found that those that had been marked touched the mark repeatedly when they saw themselves in a mirror, whereas those that had not been marked or did not have a mirror did not. Chimpanzees can also use mirrors to inspect parts of their bodies that are not normally visible to them. It therefore seems likely that the chimps recognize themselves in a mirror, and presumably, then, that they have at least some sense of self (this, of course, is still far from the usual definition of full consciousness, and is itself still a matter of debate). Some other non-human animals have also passed the mirror test, such as bonobos orang-utans, and even magpies. Interestingly, human babies less than about 18 months of age actually fail the mirror test. However, by the age of 4 they can pass fairly elaborate versions of the test, which can include elements of memory.

Human society is undoubtedly the most complex ever developed on this planet, and so it is not surprising that our consciousness is also by far the most developed. Considering just a few aspects of human consciousness makes you realize its extraordinary depth and breadth: you feel your identity and uniqueness, and yet at the same time your place in society; you have a strong sense of your character, attitudes and opinions, and a sense of morality, personal values and standards; you are aware that others have their own consciousness, and you can empathize with them; you have a highly developed capacity for language and subtle communication; you have the full gamut of emotions: love, hate, sympathy, fear, anxiety, pride, jealousy, resentment, and more; you laugh and cry; you have an ability to enjoy beauty and exercise your own sense of taste; you have a lifetime of memories; you can think in abstract terms, plan ahead and make decisions for action, often in collaboration with others; you have imagination, a capacity for beliefs, a hunger for knowledge and learning and a desire to achieve; you have the capacity to question your own existence. You can wonder at life, the universe, and your place in it.

Animals of various other species appear to share some of these characteristics, such as an ability to learn, memory, anxiety, jealousy, a sense of one's place in society, rudimentary cognitive abilities, and even, in a few cases, self awareness and a 'theory of mind'. But, even as more is learned about other species, the exceptional nature of the mental capabilities of humans will surely remain obvious, although the difference is most likely one of degree rather than absolute and qualitative.

Where's the Consciousness?

Is consciousness one single thing, or is that an illusion? The notion that there is a 'theatre of the mind', where 'consciousness happens' in the brain, including all the senses, emotions, thoughts and plans, runs into problems when we consider what we already know about the brain. The brain's activities and processing are distributed, so there would seem to be no specific central place (or time) for everything to come together.

It is known that different regions of the brain are involved with specific functions and senses. The cerebral cortex dominates the overall convoluted and crumpled appearance of the brain. It is minimal in lower animals but dominates in humans, covering the more primordial areas of the brain that control basic input and output. Many areas of the cortex have been classified according to their function and other properties. The brain is divided into two hemispheres, four lobes of the cerebral cortex in each: the frontal lobe, the parietal lobe, the occipital lobe and the temporal lobe. The two hemispheres are connected by the corpus callosum, a large bundle of nerves linking them together, essential for bilaterally coordinated cortical function. The overall brain has great complexity, which is increasingly being understood. A hundred billion individual cells (neurons) distributed throughout the brain, each with a thousand connections to the others, work together to produce all the myriad properties and outputs of what has traditionally been called the mind.

A large number of specific regions are identifiable on the basis of function, neural structure and connections. Consider the following brief overview of some of the parts of the brain, along with some of their functions. The frontal lobe is the largest division in the brain (about 40% of the cortex), and it (particularly the prefrontal cortex) contains most of the highest intellectual functions. It is the executive centre, formulating aspects of behaviour, thought, plans, judgements, wisdom and ambition, the control of movement, and language-associated memory. It defines the character of an individual more than any other part of the brain: personality control and expression, distinguishing good from bad, assessing future consequences from present actions, and generally the direction of thoughts and actions in accordance with internal input from the rest of the brain.

The parietal lobe creates the three-dimensional spatial layout of the external world and one's place in it. It is involved in attention and language. It processes sensations such as touch, pain, a sense of limb position, body orientation, and combines diverse sensations into a single experience. The left parietal lobe is involved with language-associated memory. The occipital lobe at the back of each hemisphere is primarily responsible for the processing of vision. The temporal lobe is concerned with the processing of sound and some aspects of vision, emotion, learning and memory. The primary auditory cortex, as you might imagine, is responsible for processing all auditory information, including that pertaining to language and music. Wernicke's area is very important for the comprehension of language, while Broca's area is responsible for the production of language. The angular gyrus is involved in abstraction, reading, writing, arithmetic and linking sound and meaning. The fusiform gyrus is involved with language and identifying faces.

The midbrain controls eye movements and the coordination of visual and auditory reflexes. The thalamus is a major relay point for the brain. It deals with awareness of pain, temperature, all senses except smell, and some memory processes, emotion, motivation and arousal. The hypothalamus is a major control centre, and the brain's link with the body's hormonal system, via the pituitary gland. It regulates metabolic functions, monitors information from the autonomous nervous system, and is involved in the regulation of body temperature, hunger, thirst, the circadian rhythm, emotions and sexual drive. The amygdala is mainly associated with emotions, including fear. It assesses the significance of perceptions, and is the focus of emotional stimulation and memory. It can trigger reactions such as fright, fight and flight. The dorsal anterior cingulate cortex is where anger may be triggered, and the orbitofrontal cortex is a region that helps to temper such emotional responses (interestingly, the orbitofrontal cortex is smaller in men than in women, and this may have something to do with their different inclinations to antisocial behaviour). Emotions such as happiness and sadness seem to be associated with a great many of the brain regions. Empathy is also a complex brain function, but many of the relevant functions take place in the prefrontal cortex.

The hippocampus is central in memory, in particular the encoding and retrieval of long-term memory, which can be stored throughout the brain. It is important for spatial memories and spatial navigation. In this regard it is interesting that London taxi drivers have significantly larger hippocampi than those who do not drive taxis! London taxi drivers have to do 'The Knowledge', which involves driving around the labyrinthine streets of London on a motorbike with a map on the handlebars for 6 months. Their hippocampi actually grow in the process.

The basal ganglia provide an interface between the 'higher' brain areas and the motor centres. They are important in selecting and mediating actions, and are sensitive to rewards and punishments from the outside world. The limbic system is a group of brain structures that include several of those mentioned above. The limbic structures have been described as the emotional cores of our brains, and they are central to many instincts, drives and appetites.

The list goes on. The cerebellum is involved in the coordination of movements and their timing. It regulates balance and posture, and refines the outputs of other brain areas to make them more precise. The reticular formation is involved in the arousal state of the brain, regulation of respiration and heart rates and gastrointestinal activity. The tiny pineal gland serves as a light-sensitive biological clock, regulating sleeping and waking. The brainstem is one of the oldest parts of the brain. The pons, a part of the brainstem, is an important part of the information pathway between the cerebral hemispheres and the spinal cord. It affects arousal and respiration, and plays an important role in the control of facial expressions. The medulla oblongata, a lower part of the brainstem, carries out such vital functions as digestion, breathing, heart rate, swallowing, vomiting and defecation. The spinal cord is also a major part of the brain, involved in motor control,

Memories are made of this. It's the hippocampus, deep within the brain, best known for memory formation, and whose details are being revealed by new and very sophisticated microscopic imaging techniques. Neuron bodies show up as yellow, their extensions (in particular axons) as green, and the glial (support) cells and connections are blue (Image courtesy of Thomas Deerinck at the National Center for Microscopy and Imaging Resource, University of California San Diego).

sending information to and from the brain and the peripheral nerves, and controlling respiration and heart rhythms.

Complicated as all this is, we can expect our knowledge of the brain to increase even more steeply as techniques continue to improve. A hundred years ago brain areas were identified by correlating anatomical abnormalities after death with behavioural abnormalities before death. Later, electrical stimulation during surgery provided maps. Over the past few decades powerful but non-invasive scanning techniques have been developed (positron emission tomography and functional magnetic resonance imaging, PET and fMRI respectively). Recently, maps of neural connections and a three-dimensional atlas of gene activity throughout the brain have been produced. It has even become possible to study real-time brain activity on the smallest scales by triggering individual neurons to fire. The workings of the brain can increasingly be studied in detail – and this is just the beginning.

It is clear from the above list that many of the functions we would normally consider to be part of our consciousness are performed by widely distributed regions in the brain. Many of them actually involve the coordinated participation of several brain areas at once. Might all this suggest that even the deepest and most profound elements of our 'selves' – of who we are – are functions of material processes?

Studies of the 'split brain' are of particular interest. Years ago the worst cases of epilepsy were treated by cutting the corpus callosum, the main connection between the two hemispheres of the brain, effectively separating them. Would this create two conscious selves? Experiments were informative. Visual information from the right side of the visual field goes to the left hemisphere, and that from the left visual field goes to the right hemisphere. Similarly, the two brain hemispheres control the opposite sides of the body. When patients were shown different images in the two visual fields and asked to indicate what they saw by picking from a variety of objects with their hands, the two hemispheres gave different answers, without appearing to know what the other was doing. So where is the consciousness?

Consciousness Delayed

The timing of events in the brain associated with consciousness has been much discussed over the past few decades, prompted by a series of landmark experiments by Benjamin Libet and colleagues in San Francisco in the early 1980s. These experiments revealed the famous 'half second delay', which indicates that consciousness seems to lag behind the relevant events taking place in the brain. It was first established that, following electrical stimulation of a part of the brain that caused a physical sensation, a half second passed before there was conscious awareness of the sensation.

Libet then went on to conduct a different and far more significant kind of experiment which involved timing a *voluntary* conscious decision to act. Subjects had electrodes taped to their heads and sensors on their hands. They were asked to move their hand whenever they wished, and to note the precise time of their decision to move their hand using a special clock. Three different times were measured: (1) when the subject reported the decision to move the hand, (2) the start of the associated electrical activity in the brain, and (3) when the hand was actually moved. Surprisingly, the sequence of events was in the order 2, 1, 3. The electrical activity in the brain started 300 ms *before* the subject reported the conscious decision (which in turn preceded the action itself by 200 ms).

The implication seemed clear: our conscious awareness lags behind the events in the brain that actually give rise to a decision. It appeared that the conscious intention to act does not itself cause the action. Instead, the feeling of intention, as well as the action itself, both result from subconscious processing in the brain. This immediately raised a fundamental question about free will. If unconscious processes in the brain are the real initiators of an action, then it would seem that conscious intention plays no role in the process, and free will may then be just an illusion.

These stunning results caused an uproar. Because of their importance and implications they were attacked on every conceivable front, including possible errors arising from the short time scales involved and the (unavoidable) reliance on subjective reporting, but the experiments had extensive cross-checks and controls, and have been reconfirmed over the years by subsequent experiments conducted by independent groups of researchers using increasingly sophisticated methods and technology. There is no doubt about the basic result.

More recent experiments have introduced new twists, such as having subjects decide not only when to move their hands, but also which hand to move. In an important experiment in 2008 researchers in a group led by John-Dylan Haynes at the Max-Planck-Institute for Human Cognitive and Brain Sciences in Leipzig used fMRI to map brain activity in real time, while subjects used either their left or right index finger to press a button whenever they wanted. It was found that there was preparatory activity in the frontal and parietal cortex areas fully 7 s before the subjects were aware of having made the decision to act. The sequence seems to start with unconscious decision-making, followed by the readiness potential in the supplementary motor area, then conscious awareness of the decision, and finally the action itself.

The results of this study imply not only that the timing of an action can be predicted in advance of conscious awareness, but also that the choice of *which* hand will be used to press the button can be predicted (using patterns in the comprehensive brain scans) – before the subject is consciously aware of the decision herself! And yet the subjects continue to believe that both their timing and their choice of hand were made freely. What does all this say about consciousness, and about free will?

Consciousness Disrupted

Many new facts have come to us serendipitously via patients who have brain damage or abnormalities of various types. These patients can be studied by neuroscientists using nonintrusive techniques – clever experiments dealing with response and behaviour, increasingly sophisticated brain imaging, etc. A few examples are given below.

One is face blindness, caused by damage to the temporal lobes on both sides of the brain. The symptom is that the patient can no longer recognize people just by looking at their faces, even though the patient can still see and read, and appears normal in other respects. Another example is the well-known phenomenon of phantom limbs. A patient missing a limb because of an accident or amputation may sometimes feel sensations 'in that limb'. There is an extraordinary condition called synesthesia. It is an amazing mixing of the senses: musical notes and visually perceived numbers can give rise to a sense of colour (middle C may invoke the colour green, 4 may be seen as orange). Certain types of damage to the parietal lobe may result in colour blindness (the world is seen in shades of grey), or in other cases an inability to sense the direction or speed of a motion.

Other examples come from some of the most prevalent and well-known mental diseases. Alzheimer's disease is the most common form of dementia. The first major symptom is short-term memory loss, and symptoms in advanced stages include confusion, mood swings, loss of long-term as well as short-term memory, language difficulties, problems with executive functions and abstract thinking, delusion, and decreased motor coordination. In late stages extreme apathy and exhaustion set in. The sceptic may argue that these are outward symptoms that don't necessarily have anything to do with consciousness, which is supposed to be internal and private, but for most of us it is hard to avoid the conclusion that 'confusion', 'mood swings', and 'delusion' must indeed reflect the 'inner state'. Striking physical changes accompany these symptoms, including extreme shrinkage of the cerebral cortex and hippocampus, and the presence of plaques around the neurons and 'tangles' inside the neurons. The close relationship between physical damage and the mental symptoms is obvious.

Another severe type of brain damage showing specific links between cause and symptoms is stroke. This is the third-largest cause of death in the world. It is due to a problem in the blood supply to the brain, caused by either a blocked or a burst blood vessel. The onset of symptoms is immediate, and they are clearly linked to the part of the brain that is affected. If it is the cerebral cortex, symptoms can include problems related to language, memory, seeing, voluntary movements, thinking and confusion. If it is the cranial nerves of the brainstem, some of the problems may be related to the senses, reflexes, and breathing. If it is the cerebellum, there may be problems with movement, coordination and equilibrium. And, as may be expected, the broader the area of the brain affected, the more functions that may be lost. The tight relationship between the specific part of the brain affected and the symptom is of course what would be expected from the detailed brain studies summarized above.

Parkinson's disease is readily identifiable from a person's movement disorders. What is less well known is that there can be secondary problems related to problem solving, memory, speech, and mood alterations. The cause of the symptoms is a deficit of the neurotransmitter dopamine. The decreased dopamine activity can readily be seen in brain scans.

Over the past century, some radically invasive forms of brain surgery were carried out as last-ditch efforts to deal with extreme forms of mental illness. A lobotomy, for example, involved cutting the connections to and from the prefrontal cortex. The result was usually the blunting of the personality and apathy. Another example, mentioned above, is cutting the corpus callosum in order to stop epileptic seizures. The result is obviously a split brain: communication between the two hemispheres is much reduced or impossible, and bizarre behaviour can result. Such drastic operations show dramatically how much the material brain can affect conscious behaviour.

Consciousness can certainly be disrupted by anesthesiology. There are two extremes, total wakefulness and total anesthesia, with a continuous spectrum in between. There are certain operations, such as a cesarean delivery, in which conscious sedation is applied to provide comfort while permitting the mother to participate in the birth. There are cases in which the patient can respond to commands during the operation and yet does not have any recall afterwards, and vice versa, depending on the dosage. Anesthesiology is an amazing and hugely successful practice, and it certainly demonstrates the power of drugs over consciousness.

Conscious or Unconscious?

We are normally unaware of just how much of our daily activity is carried out unconsciously. When we turn a doorknob or climb stairs, we are certainly not aware of the detailed motor commands from the brain and the myriad muscle movements required for the task – we do these things unconsciously. If someone slips when walking beside you, your hand immediately goes out to help them, with no consciousness contemplation. We learn how to ride a bicycle, but from then on it's pretty much 'second nature', and our consciousness doesn't have to be bothered with all the twists and turns. Another commonly cited example is driving a car. After you've driven 5 or 10 minute to the supermarket, talking to your passenger all the way, can you remember anything at all about the drive itself? You can certainly remember the discussion you had. Can it be said that you were conscious of one (the discussion) but not the other (the driving)?

Think of the basic human instincts listed in Chap. 14, such as 'fright and flight', eating, chewing, digesting and excreting. Many of these we have in common with 'lower' animals that are considered to be automata. Again, our brains take care of these things for us, leaving our consciousness free for grander things.

There are significant genetic components in many of our traits and abilities, again as outlined in Chap. 14, even including things as fundamental as a sense of fairness. These influence what we do and how we do it. Whether the course of action is decided consciously or unconsciously, the decision and the subsequent action are influenced by what is written in our genomes and the associated unconscious processes in the brain.

It's worthwhile to take a bit of time and effort to be aware of your own conscious and unconscious behaviour. As an observer of yourself, you can learn quite a lot. Try separating out the various simultaneous experiences you are having – the sights, the sounds, the taste, your own thoughts. It's actually not too hard to do. What fraction of your actions are 'involuntary body functions'? Was the decision to take a particular action made consciously or unconsciously? Does this depend on the length of time leading up to the action (avoiding a falling rock, or planning next year's trip)? Was the execution of the action conscious or not?

Psychologists have studied how actions can be influenced by preparatory stimuli with intensities too low for conscious awareness ('subliminal stimulation'). Such stimuli are often in the form of words related to achievement, assistance, reward, generosity, physical exertion and the like. It is found that such subliminal priming significantly influences goals and how they are carried out. The priming is also (significantly) found to enhance brain activity in regions related to the specified goal. When the subjects are later asked about their conscious awareness of the motivation, it is found that they really were unaware of the subliminal priming. The influence of the priming on both the decision to act and the action itself is apparently unconscious, indicating that preparing and achieving a goal can happen without conscious awareness. It seems that behaviour can sometimes originate in an 'unconscious will'. The brain can, as they say, be steps ahead of 'its owner'. This has significant implications for our normal feeling that we have free will and can consciously decide every-thing we do.

Finally, as noted above, human babies fail the mirror test for self-awareness, which some non-human animals pass. It takes time for consciousness to develop. What does this say about our lifelong sense of self-awareness and continuity? At the very least, it appears that what we normally consider to be our special human consciousness overlaps to some extent with the capabilities of some other animals. There appears to be a continuum, although the development of the human brain and consciousness was obvi-ously a huge quantitative step.

The Nature of Consciousness

What causes this phenomenon we call consciousness – this private view of the world, this overwhelming and all-encompassing sense of self, this acute awareness of sensations and emotions, this continuity of memories and plans, this source of judgement and morality, this comprehensive world view?

Such questions long ago naturally gave rise to the notion of dualism – that consciousness and the brain are materially differ-ent from each other. In this view, the consciousness we experi-ence is literally supernatural – it is not part of the natural world. But dualism does not explain what consciousness actually is, nor does it explain how a supernatural consciousness – by definition totally separate from the natural world – could cause events in the material world, which includes the brain. Dualism does not exist in modern scientific thinking about consciousness.

The broad consensus in the scientific community is that consciousness is a function of the material brain. John Driver, Patrick Haggard and Tim Shallice of University College London summed it up in a Royal Society lecture in 2007, in which they said: "Advances in neuroscience have now led to wide acceptance in science and medicine that *all* aspects of our mental life – our perceptions, thoughts, memories, actions, plans, language, understanding of others and so on – in fact depend upon brain function." We have seen above many examples in which individual elements we would normally associate with consciousness are clearly related to material processes in the brain. We have seen that consciousness sometimes appears to follow after events rather than preceding them (it appears to keep track of what is going on, but may sometimes be too slow to actually cause things to happen). And we have seen how consciousness is affected by traumas, split brain experiments, and brain diseases. In just the last decade there have been many technological and methodological breakthroughs (especially neuroimaging). It certainly appears that consciousness is a product of the brain – ultimately of neurons and chemicals. It does not appear to be a local activity, but instead seems to involve a large variety of sites across the brain. So much so that one could say that the original question, "Can we identify the 'neural correlates' of consciousness", is now being replaced by the opposite question, "How do the diverse and myriad processes of the brain somehow come together in the end to create the sensation of one single comprehensive consciousness?"

A Few Implications

If consciousness is purely a function of the material brain, there are some fundamental implications. First, just as the concept of the life-force, which was supposed to distinguish living matter from inanimate matter, became obsolete when the mechanisms of life were understood, so the mystery surrounding consciousness may fade away when the workings of the brain are better understood. Second, consciousness would not exist without the brain. Death of the material body, including the brain, would also be the death of any dependent consciousness. Any consciousness which is based on material processes would clearly cease to exist when they do.

A third implication has to do with the age-old question of free will. If consciousness is a part of the physical world, then, from a strictly deterministic point of view, it is as subject to the laws of causality as anything else. This means that every action we make, every decision we take, is pre-determined. However much it may feel as if we have free will, this is inescapable. According to the strict determinism of classical physics, there is a (perhaps infinite) chain of cause and effect, and everything, to the smallest detail, is absolutely pre-determined by what has gone before.

It has sometimes been suggested that this chain can be interrupted by chaotic events. The weather provides examples of chaotic events: conditions develop exponentially into extremes, producing sudden and catastrophic storms, after which the weather calms back down again. Even our most powerful computers can't follow the complexity of such chaos. But that doesn't mean that cause and effect have broken down. Causality at all levels, from atoms to clouds, remains valid, even if we have trouble following it ourselves. Ultimately it is still straightforward classical physics.

Another form of determinism has arisen from the development of genetics: biological determinism. According to this view, what we are is determined entirely by our genomes. Aside from the fact that the environment can play a significant role in the expression of the genome, the very concept of biological determinism would seem to beg the question of what chain of prior events determined our genomes. It would seem rather arbitrary to pick out any particular part of a causal chain (in this case our genomes) as being essentially the 'first cause', ignoring all preceding events and considering only those that followed.

However, quantum uncertainty, which became known in the last century, is a far more serious issue, which does imply a fundamental limitation on determinism. At the level of quantum mechanics, events can only be predicted in terms of probabilities, even in principle. The absolute certainty of classical physics no longer applies, and neither, presumably, does the determinism that it implied. It has been argued that quantum uncertainty only applies on subatomic scales, and not on the vastly larger biological scales, but how can there be a clear dividing line on any scale? Some have claimed that quantum effects can indeed play a significant role in the processes of the brain, although this is widely doubted. But even if that were true, it would just replace determinism with its complete opposite – total randomness – which is also not exactly how most people would like to consider their free will. These issues remain subjects of debate, leaving the status of physical determinism in limbo.

But these discussions of determinism are dwarfed by the complexities of the issue of free will itself. It is even debated whether or not determinism has anything whatsoever to do with free will. Incompatibilism holds that free will is not compatible with determinism, while compatibilism maintains that it is. There is a wide spectrum of views under both of these headings and many others as well, involving moral, social, legal and other such issues. Free will is the ultimate philosophical chestnut, having been debated for millennia. Who knows, to clear the air and remove the baggage associated with human social issues, perhaps we may ultimately best understand determinism and free will by studying other animals!

But determinism is just one aspect of the discussion on free will. As we saw above, solid experimental evidence related to free will is now available, thanks to the work of Benjamin Libet and others. The conscious intention to act, it seems, does not itself always cause the action. Rather, both it, and the action, can follow from previous unconscious activities in the brain. The findings of psychologists on 'unconscious will' indicate something similar. It will require fancy intellectual footwork to avoid the conclusion that free will as normally conceived is an illusion – but never underestimate the subtle inventiveness of philosophers.

16. Are We Special?

To ourselves, we are of course very special. But is there any objective sense in which we could be said to be special?

Do We Have a Privileged Place in the Universe?

As long ago as 500 BC the ancient Greeks were speculating about the universe and our place in it. Bold ideas were contemplated, including some that are established parts of modern science today. It was suggested that the Moon has mountains, that the Earth moves around the Sun, and that the universe contains vast numbers of stars. Some even contemplated that other worlds may exist that contain life. But there were many other and conflicting ideas in that world of free thinking.

One enduring legacy was a geocentric theory, formalized by Ptolemy, describing the motions of the Moon, Sun and planets about the Earth. This was a true scientific theory insofar as it made clear predictions that could be tested by observations. And tested it was. Over the centuries it had to be corrected and elaborated so extensively by adding 'epicycles upon epicycles' to describe the motions that it became unbearably cumbersome. The fundamental problem was the assumption that the Earth was at the centre.

A far simpler theory placing the Sun at the centre was proposed in 1543 by Polish astronomer Nicolaus Copernicus. This was a monumental advance that changed science, and is referred to as the Copernican Revolution. It was not a painless revolution. In 1600 Giordano Bruno was burned at the stake by the Roman Inquisition, in part for being a proponent of this heliocentric model. A more famous Italian, Galileo Galilei, was condemned to lifelong house arrest for the same crime. Galileo had used the newly developed telescope to see mountains on the Moon, discover four moons orbiting Jupiter, observe the phases of Venus, and resolve countless distant stars like the Sun. These historic discoveries provided strong, direct support for the heliocentric view. Also in the early 1600s, Johannes Kepler discovered the three laws that explain the motions of all the planets around the Sun (ellipses, not circles). Later in the same century, Isaac Newton derived Kepler's laws using his new theory of universal gravitation. The Copernican Revolution was complete; the Earth is not at the centre of the solar system.

Well, perhaps our solar system is special. What about the Sun itself? No, it turns out that our Sun is a typical star, one of a hundred billion in our galaxy. Since the late 1800s we've been able to study the properties of our Sun and countless other stars using spectroscopy. We now understand stars very well. There is a 'family' of stellar types, and our Sun is a typical star in the middle of the distribution.

Are the Earth and the other planets in our solar system the only ones in the universe? Is our solar system unique? No, we have now observed over 500 planets orbiting other stars, and there may be billions in our galaxy, many of which may harbour life.

Is our solar system located in a special place in the galaxy? No, we have studied the structure of our galaxy extensively, and we know that our solar system is located 27,000 light-years from the centre of our galaxy – a medium distance, not too close and not too far. It is also located between two of the spiral arms of our galaxy – not a very special place (the spiral arms are where the exciting things such as star formation take place, and they contain many of the exotic phenomena).

Is our galaxy special in any way? No again. Of the billions of galaxies in the visible universe, ours is pretty ordinary. It is a spiral galaxy, one of the three broad classes in the 'local' universe originally described by Edwin Hubble and others. The amount of light emitted by our galaxy also places it in the middle of the range. Hubble's discovery that virtually all galaxies are moving away from ours does not mean that our galaxy is at the centre of the universe. Recall the model (described in Chap. 2) of the two-dimensional dots on the surface of an expanding three-dimensional balloon: every dot sees all other dots as moving away from itself, but there is actually no centre. No galaxy is special in this regard. Our galaxy resides in a small group of galaxies, which is gradually falling into the relatively nearby Virgo cluster of galaxies due to gravitational attraction (on a timescale of billions of years). Galaxies move around all the time under the gravitational influence of other galaxies, and this is nothing special.

What about the overall composition of the universe? As described in Chap. 3, the mass-energy of the universe is comprised of dark energy (72.1%), exotic dark matter (23.3%) and 'ordinary' (atomic) matter (4.6%). We only understand the 'ordinary' matter – we do not know what the other 95.4% is. We (and life as we know it) are made of ordinary matter. Some have dubbed this situation 'the ultimate Copernican Principle' – we are not even made of the dominant stuff of the universe.

Finally, our universe may not be the only one. As mentioned in Chap. 5, speculations over the last decades have raised the possibility that there many be many universes, perhaps an infinity of them, each of which may have different laws of physics. If that were true, then it might seem to diminish our significance even further.

So we really cannot consider ourselves to be special on the basis of our place in the universe. This situation has also been given a name – 'the Principle of Mediocrity'.

Our Place Here at Home

On the other hand, there is no question that we humans presently occupy a dominant place on Earth. Of course all other species are special and unique in their own ways. They can fly, see ultraviolet light or the polarization of the sky, echolocate and live in the sea; we have no natural ability to do any of those things. Evolution guarantees that every species is exquisitely well fitted to its environmental niche, with the expertise required for that niche. Nevertheless, we have certainly become dominant.

So did many other animals, in their time. The dinosaurs dominated in an obvious way for over a hundred million years, at least amongst the macro-animals. Each epoch in the Earth's history must have had a dominant species in one sense or another.

Numbers don't tell the whole story, but they can be interesting. The human population is approaching a staggering seven billion. We humans are many thousands of times more numerous than non-domesticated mammals of comparable size. Very roughly, as these numbers are hard to determine with any precision, the ratios of the populations of lions, tigers and wolves relative to that of humans are only about 0.000003, 0.000001, and 0.00003 respectively. Even for our close relatives the chimpanzees the ratio is less than 0.00003. Cows, sheep, dogs and cats are considerably better off, at about 0.23, 0.17, 0.05 and 0.05 respectively (clearly, domestication has benefited many animals, at least in this restricted sense; dogs, which evolved from grey wolves, are about 2,000 times more populous than their wild cousins). We are certainly the most numerous of the large land-based creatures on Earth. So, numerically at least, we have done very well.

We are full members of the tree of life (although we account for only a small fraction of the total biomass of the Earth). We are just one of many millions of species that share exactly the same genetic code, the same cells and the same mechanism for metabolism – amoebae, insects, fish, flowers, trees, cows, and all the rest. There is also a continuity across the family of life in development, from embryos to cognition. There is no gap that separates us from other species. Here again, we are not special.

But then how did we, with our humble beginnings, achieve dominance not just numerically but in so many other ways? Several factors have combined to make this possible. Our big brains are obviously by far the most important. They have enabled us to do far, far more than required for living on the savannahs of Africa. We developed weapons, tools and clothes (essential for expanding our habitat), and we domesticated fire. Language has given us an enormous advantage. Any clever invention, like the wheel, only has to happen once; the idea is then transferred by language to the rest of humanity. We now live in an age of astonishing scientific and technical progress, and the benefits belong to all of us, through language. The development of language required our big brains, flexibility in our throats and mouths, and good hearing backed up by impressive computing power. The opposable thumb was essential for the dexterity we now have with our hands, as was the rotational flexibility of our forearms. Being able to stand and walk on two feet was essential so that our hands were free for their many other uses. Binocular vision was important in connection with the use of our hands.

None of these features by themselves are unique to humans. Many other animals can walk on two limbs, many have opposable thumbs (including most primates), and many have binocular vision. Aside from our big brains, perhaps the feature that may have developed more in humans than in other species is the larynx (the 'voice box'), which generates sound. The importance of the larynx in the development of speech and language is debated, but there can be no question as to the sophisticated degree of sound control made possible by the human larynx and tongue. There is some evidence that adaptations for speech may have developed in our human ancestors sometime over the period between 0.5–1.5 million years ago. Were these refinements to the larynx essential for language, or only beneficial? After all, we can also communicate using sign language, writing, and even using the oesophagus, although it is hard to imagine that vocal speech using the larynx wouldn't have become the preferred mode.

These diverse features have amazingly come together at the same time in one species – humans – and have all been essential for our dominant place in the world today. Our intelligence enables us to copy the special advantages that members of other species were born with (ultraviolet vision, flying, etc.). But we can do far more – consider radio communication, travelling to the Moon, and many other achievements that no other species could ever come close to. Now we can even manipulate the very genetics of any species we choose (including our own).

But we remain vulnerable. A large meteorite impact could kill us as surely as the dinosaurs, while leaving much of the biomass – the millions of species that live under the surface of the Earth – unscathed. Our very achievements could be our downfall. Nuclear holocaust or biological disaster could wipe humanity off the face of the planet. Nevertheless, at the moment, we are in an extraordinary position.

In summary, we certainly do have a remarkable combination of abilities that has enabled us to become the dominant species on this planet, although we do not seem to be special in any other way.

Are we as good as it gets? We are only comparing ourselves with other creatures here on Earth. Are we at the top end of an absolute, universal scale in brain power, which, even in principle, can go no further? That seems highly unlikely. We ourselves continue to evolve. Who knows how much further that scale may extend? How do we compare with other life that may be present elsewhere in the universe? Indeed, *is* there any other life in the universe?

17. Are We Alone in the Universe?

The universe may well be teeming with life, while we remain blissfully unaware of it. In our own solar system there may be life forms similar to some of the extreme varieties we now find on Earth located on other planets or moons. If even a small percentage of the hundred billion stars in our galaxy has planetary systems, the odds for the presence of life elsewhere in the galaxy may be huge. Interstellar space contains complex molecules, including amino acids, some of the building blocks of life, and these may be delivered by comets and meteorites to nascent planets. In that case life may have a head-start even before planets are formed. And there is the possibility of panspermia, in which primitive life forms may be transported from one planet to another. And then there are the billions of other galaxies in the universe. It seems hard to imagine that we are alone.

The Slippery Definition of Life

If someday we were to discover possible evidence of life beyond the Earth, it would be most helpful if we had agreed beforehand on the definition of life. However, as pointed out earlier in this book, this is not as easy as it sounds. There seem to be exceptions to almost any suggested list of defining criteria: inanimate phenomena that would have to be included in the definition of life (like fire), or living creatures that would have to be excluded (like mules).

Autonomy, metabolism and replication are perhaps some of the most useful characteristics for a definition of life. Autonomy would imply a well-defined distinction from the surroundings. Metabolism would require an appropriate source of energy, which may be detectable in the environment. Evidence of

P. Shaver, *Cosmic Heritage*, DOI 10.1007/978-3-642-20261-2_17,

replication would probably be far more difficult to pin down. Evolution, which has often been suggested as one of the defining characteristics of life, follows naturally from imperfect replication in the presence of a challenging environment, and so should probably not be included as a separate item in the definition. Are there other characteristics that should be included? This is important, as we should be prepared for any form of life, not just life as we know it here on Earth.

What Should We Expect?

What are the requirements for life? Certainly we would all agree that life requires an energy source of some kind. Life as we know it here on Earth uses direct sunlight to produce energy through the process of photosynthesis, or (especially deep under the surface of the Earth) chemical reactions involving either organic or inorganic molecules. These are abundant sources of energy on our planet. Elsewhere they may not be so abundant, but alternatives may conceivably provide the required energy. Possibilities that have been discussed include stellar radiation outside our conventional optical window, thermal (and other) gradients, tides and tidal flexing, tectonic effects, convection, radioactivity, magnetic fields, and several others. Determining which of these may be important requires a careful study of the physics and chemistry in each case.

Life as we know it is based on carbon. Is this the only possibility? Could there be other forms of life with a totally different chemical basis? Carbon does have a number of major advantages. It is unique in its ability to form a vast range of stable complex molecules together with other elements in a wide variety of shapes – long-chain polymers, rings, and three-dimensional macromolecules. This is possible because a carbon atom can simultaneously bond with up to four other atoms. Such characteristics provide enormous richness and complexity to biological systems. The carbon–carbon bond energy is exceptionally strong. Carbon also bonds strongly to both hydrogen and oxygen, and is in general highly compatible with water.

Of possible alternatives to carbon, silicon is the most promising. Its physical properties are quite similar to those of carbon. It is the only other element that can form four bonds at once, although they are relatively weak. Therefore, like carbon it is versatile, and can also form long molecular chains. The compounds it can form would only be possible under restricted environmental conditions: temperatures and pressures outside of the typical range on Earth, the presence of a liquid such as methane in place of water, and an absence of oxygen and carbon. Other elements have also been considered (such as boron, nitrogen, phosphorus and sulphur), but most if not all appear unsuitable as the basis for a living system. It seems that carbon is by far the most likely basis for life, except in what we would consider to be extreme environments.

Water is often mentioned as the key to finding life. Why? Because in life as we know it, a liquid medium in a cell is required for molecules to move freely and interact, making possible the chemical reactions that led to life and now sustain it (the human body is two-thirds water). It can facilitate and take part in such reactions. Water also provides a relatively stable environment, with upper and lower bounds to fluctuations in temperature. It is well suited to the conditions prevailing on the Earth, and has other beneficial properties, such as its bipolar structure. It indirectly provides the ozone in the atmosphere that shields the Earth from harmful ultraviolet radiation (photosynthesis releases oxygen, which in turn produces ozone by photochemical processing in the atmosphere). And the elements that comprise water are the most common in the universe (hydrogen) and on Earth (oxygen). Water is ideal for life on Earth, and is adequate over a wide range of conditions.

However, on planets with very different temperatures, densities and other characteristics, other liquids may be preferable, such as ammonia, methane, sulphuric acid, methanol, sulphur dioxide, and hydrocyanic acid. Some liquids may be better than others at different depths within the same planet. The search for life will ultimately have to be wide.

Where is life most likely to be found? On the surfaces of planets? Beneath the surfaces? In planetary atmospheres? In space? We are of course most familiar with life on a planetary surface. Here we have plenty of sunshine for energy, and a large and varied surface on which to expand and develop. It's hard to imagine living in any other way. But there are serious hazards. We are exposed to extremes of weather, as we are reminded almost daily. Earthquakes and their associated tsunamis, massive volcanic eruptions, and rare events such as impacts by large meteorites can be totally devastating. Major climatic changes can have long-term catastrophic consequences. Life on a planetary surface is vulnerable.

By contrast, life beneath the surface can be stable and secure over the long term. The meteorite that killed off the dinosaurs probably had minimal impact on subsurface life. Far beneath the surface the temperature is stable, there is little or no risk from damaging solar radiation, tsunamis, landslides, volcanoes, ice ages, or meteorites, and there is a plentiful supply of chemical energy. The downside is that most if not all subsurface life is likely to be and remain microscopic.

Life within a planetary atmosphere may seem unlikely, but perhaps not if the atmosphere is sufficiently dense and contains suitable chemicals. Some significant organic molecules may have formed in primordial atmospheres. Any living systems in a planetary atmosphere would probably be microscopic, for reasons of buoyancy. The best candidate for atmospheric life in our solar system may be Venus.

Interplanetary or interstellar space is almost certainly an impossible home for living systems, on account of the homogeneity of space, the extremely low temperatures and pressures, and the hostile radiation environment. However it may be possible for 'transients' to survive an extended trip through space: dormant endospores protected from radiation by surrounding rock could likely survive for long periods of time, so the idea of panspermia, at least within individual planetary systems, may be feasible.

Have we been too conservative in contemplating what forms of life there may be in the universe? As mentioned earlier in this book, it has been said that "whatever is not strictly forbidden (in the universe) is mandatory." Nature will find a way of doing it. We have often been surprised by what we have discovered in the universe. Extremophiles provide an example right under our noses, showing that we should expect the unexpected. Their discovery took us by surprise in the early 1970s, even though they live right here on Earth and comprise a significant fraction of the

Earth's biomass. They can live in extremes of both heat (in the superheated waters of Yellowstone) and cold (in glaciers of the Antarctic). They thrive in the hot vents kilometres deep in the ocean. They live in rock kilometres below the surface of the Earth. They are still members of our carbon-based family of life, but they live in conditions we thought were totally impossible. Having found such extremes of life here on Earth, think of what we might find elsewhere in the solar system and beyond. The ultimate exotic life forms would be those engineered and made artificially by other intelligent creatures living elsewhere in the universe. In that case, all bets are off as to what we might find.

Why are we so obsessed with carbon–water forms of life? The carbon–water combination does have many advantages in our universe, as outlined above. But in addition, it is obviously the easiest for us to recognize, as it is the only one we know. If we are going to design expensive space missions and telescopes to look for life, it may be most cost-effective to search for life as we know it. Of course we then run the risk of missing other (and to us exotic) life forms. We can judge these trade-offs below when we consider strategies in the search for life.

Signatures of Life

We would ideally like to find extant extraterrestrial life itself (and searches continue), but short of that, what would be the best *indirect* indicator that such life does (or did) exist? Undoubtedly the best would be the discovery of unambiguous fossils. Such a claim was made in 1996 based on a Martian meteorite found in the Antarctic. The original rock of the meteorite is about 4.6 billion years old, the age of the solar system itself, and was dislodged from Mars by an impact some 16 million years ago. After 13,000 years in space the meteorite was caught by the Earth's gravity and plunged into the Antarctic ice. The case that it may have contained life was based on the presence of structures reminiscent of fossilized micro-organisms, organic compounds, and the possible products of organic activity. The claim was immediately surrounded in controversy, and remains ambiguous today.

Aside from fossils and probes for evidence of (past) microbial activity, we are left with remote sensing, and many possible signatures have been suggested. Vast quantities of organisms can together produce direct evidence of biological activity. Indirectly, organic activity on a sufficient scale can actually modify the environment, leaving possible signatures of life. Examples include ozone or molecular oxygen in an atmosphere, various other molecules consistent with life processes, large quantities of limestone, high rates of erosion, obvious structures such as coral reefs, unusual chemical ratios, the presence of macromolecules and chirality (left- or right-handedness) in their structure, evidence of disequilibrium in the physical or chemical environment (e.g. unusual heat or chemical gradients), the presence of a liquid medium, a dense atmosphere – and there are many others. Of course the range of possibilities explodes if we include the conceivable effects and artefacts of intelligent life.

Life in Our Solar System?

The search for life in our solar system begins right here on Earth. We certainly know of millions of species, but they are all part of one family, based on carbon and water. The interesting question is whether there could be totally different forms of life here in our own back yard – perhaps life based on silicon (after all, our relatives, the extremophiles, were a total surprise to us). There are at least two reasons for thinking the answer may be no: (1) we have never found any, and (2) even if they once existed, in the scramble for resources they may have been totally outcompeted in all possible niches by our own carbon–water form of life. An interesting bacterium is known that lives in an 'arsenic-rich' lake, but whether it actually uses arsenic (instead of the usual phosphorus) in its key functioning molecules remains to be determined. It would, of course, be wonderful if we were ever to prove that alternative forms of life exist, as we would learn so much more about the possibilities for life everywhere.

The Moon has for decades been considered an unlikely place for life. It has no atmosphere, and is therefore totally exposed to the hostile radiation and bombardments from space. Temperatures are extreme, averaging above 100°C during the lunar day and below −150°C during the lunar night. However, a recent man-made explosive impact has revealed the presence of water ice beneath the surface of a permanently shadowed crater at the lunar south pole, in addition to possible traces of other molecules such as carbon dioxide and methane. The overall amount and distribution of the lunar water ice are unknown at present, but its very existence may have significant implications. Even if the Moon still turns out to be unpromising for 'natural' life, it may conceivably provide a self-sustaining permanent base both for human colonization and for explorations further into the solar system.

Mars is a special place for the search for extraterrestrial life, as it is (along with the Moon and Venus) the only place where direct sampling is readily feasible; it is as little as 6 months' travel from Earth. In many ways it is somewhat similar to the Earth. It has an atmosphere, albeit thin and mostly comprised of carbon dioxide. It is cold, with an average temperature of about −60°C. Water ice is present at the polar caps, and has been detected elsewhere near the surface of the planet. There is evidence that liquid water once flowed on the surface, causing many of the physical features visible today. There is also considerable evidence for the presence of liquid water beneath the surface. In that case there could be subsurface life.

There have been several robotic missions to Mars, most prominently the Viking missions of the 1970s, and dozens of Martian meteorites have been found on the Earth. In all cases so far (including the famous Antarctic meteorite mentioned above) there has been no unambiguous detection of Martian life, but the prospects for finding it still seem good. Several more missions are planned, taking advantage of our much better understanding of Mars, and manned exploration may follow.

Venus is enshrouded in a very thick atmosphere, again almost all carbon dioxide. The atmospheric pressure is huge, almost a hundred times that on Earth. The abundant carbon dioxide causes a 'greenhouse effect', producing an extreme temperature of 470°C on the surface of the planet. However, depending on the past history of Venus, there is one place where life could conceivably now exist: in the clouds of the thick atmosphere, where the temperature is far lower than at the surface. Future robotic missions may be able to return samples from this region of the atmosphere. The innermost planet, Mercury, is thought to be unpromising for life because of the strong solar radiation, the lack of an atmosphere, the extremes in temperature, and the likely absence of liquid water. Unlike the terrestrial planets, the four outer Jovian planets – Jupiter, Saturn, Uranus and Neptune – are unlikely hosts for life because they are all gas giants, and lack any solid surfaces or structures on which life could have developed.

But several of the *moons* of the Jovian planets may be more promising for life – in particular Europa and Ganymede (moons of Jupiter) and Titan (a moon of Saturn). Like Mars, Europa may have a subsurface liquid water ocean, and hydrothermal vents like those at the bottom of Earth's oceans. A dedicated unmanned orbiter could be sent to confirm the presence of the subsurface ocean. Ganymede, the largest moon in the solar system, may also have a subsurface ocean. Titan has an atmosphere that is rich in organic chemistry, and it has surface lakes of liquid ethane and methane. It could conceivably support interesting and varied forms of life.

Thus, there are many possibilities for the search for extraterrestrial life within our solar system – a search that will undoubtedly escalate over the coming decades.

Life Beyond Our Solar System?

There is an obvious sequence of questions to ask in exploring the possibility of life beyond our own solar system. (1) Are there other planets in the universe, aside from those in our solar system ('extrasolar planets')? (2) Can any of them support life? (3) Can we find indirect evidence for extraterrestrial life? (4) Can we find extraterrestrial life itself? (5) Can we even find extraterrestrial intelligence?

In one of the great astronomical discoveries of the past century, the first of these fundamental questions has already been definitively answered. Astronomers had long believed that there must be planets around others stars (this idea goes back at least to the time of the Greek philosophers over 2,000 years ago), but until they were actually detected this was just speculation. Now we know.

In 1992 radio astronomer Alexander Wolszczan discovered unusual systematic variations in the timing of a pulsar (a rapidly rotating collapsed star). Pulsars are near-perfect timing devices, in some cases as good as atomic clocks. The variations could therefore be measured to incredible precision. They were found to be consistent with the gravitational tugs of two planets orbiting the collapsed star. This was quickly established beyond doubt. However, as the collapse of the star that formed the pulsar occurred as part of a supernova explosion, the two planets would have been rapidly fried into cinders. A system of two planet-mass cinders orbiting a pulsar was thought by many to be too exotic to be considered a *bona fide* case for extrasolar planets. But there is no question – these are two planet-mass objects orbiting a (dead) star.

How do we find an ordinary planet orbiting an ordinary star? The very faint light we would see from the planet is just reflected light from the star – a million times fainter than the light from the star itself. Imagine trying to see a tiny insect flying beside an extremely powerful spotlight that is pointed straight at you in the dead of night. There must be a better way.

A bit of lateral thinking led to the first successful method: look at the star, not the planet. The star is not completely stationary. It is pulled and tugged *very slightly* by the gravitational attraction of the planet. As the two go around each other, the star is pulled towards us when the planet is on the near side, and pulled away from us when the planet is on the far side. The periodic backward and forward motion of the star causes the absorption lines in the star's spectrum to shift slightly back and forth across the spectrum, due to the 'Doppler effect' described in Chap. 1. These shifts can be observed using highly accurate spectroscopy, giving a direct measure of the motion of the star.

In 1995, Swiss astronomers Michel Mayor and Didier Queloz used this technique at the Observatoire de Haute Provence to discover the first planet orbiting another ordinary star (called 51 Peg). This was rapidly followed by a burst of such discoveries, and now, just 16 years later, over 500 of these 'extrasolar' planets are known. Within the next decade thousands will be known. Our solar system is not unique in the universe.

The planet found by Mayor and Queloz is certainly weird. It is a Jupiter-mass planet orbiting so close to its star that its orbital period is just 4 days! By contrast, the Jupiter we know in our own solar system has an orbital period of 12 years, and an orbital radius 100 times larger than that of the 51 Peg system. The newly discovered planets of this type are understandably called 'hot Jupiters'. They were a total surprise to astronomers. How do they manage to survive, and not break up? How could something the size of Jupiter have got that close to its star in the first place? Current theories focus on some sort of migration process. In any case, it's no wonder that they were the first to be discovered: because of their small periods and relatively large gravitational effect on their parent stars they were the easiest to detect. That most of the first extrasolar planets to be discovered are hot Jupiters is an observational 'selection effect'. It's harder to find planets with lower masses and/or larger orbits, so that is where most of the current work is concentrated. Much better spectrographs, bigger telescopes and longer observation periods are being used, and have now succeeded in finding planets just a few times more massive than the Earth itself.

Other methods have also been used since that first discovery. One obvious one is the so-called transit method. If the orbital plane of the star-planet system happens to be aligned fairly edge-on relative to our line of sight, then the planet will periodically pass 'in front of' the star (i.e. between the star and us). Each of these transits will be a partial eclipse, causing the light we observe from the star to be diminished slightly during the length of the transit. The first such transiting event was observed by David Charbonneau and colleagues using a 10-cm amateur telescope on a parking lot in Boulder, Colorado (you can sometimes do ground-breaking science with modest facilities!). The same system was observed a few months later using the HST, and the quality of the 'eclipse' data was astounding. Two other telescopes in space are now making observations of this type: the COROT satellite launched by Europeans for a different purpose, and NASA's new Kepler spacecraft which is dedicated to this project. Kepler will find many hundreds of transiting extrasolar planets over its 4-year mission; it has already found over a thousand candidates.

A different kind of transit method has also been used. In this case it involves the planet going *behind* the star and reappearing. The planet is as bright as it will ever get (as seen by us) just as it is about to disappear behind the star and just after it has reappeared from behind the star. This is because, in these cases, the planet is fully illuminated by the star, so we get the maximum reflected light. To help overcome the 'spotlight' problem mentioned above, the observations are made in infrared light, because the planet is warm and radiates more at these wavelengths, whereas the star is not as bright in the infrared as it is in the optical. For both reasons the ratio of the planet's brightness relative to that of the star is increased (relative to what it would be at optical wavelengths and in other configurations), making detection a bit easier. The technique is simply to observe the brightness of the total system: it should be slightly brighter just as the planet starts to go behind the star, then slightly dimmer because the light is now only that from the star itself, and then slightly brighter again as the planet emerges from behind the star. This effect has now been observed in several cases using NASA's very sensitive Spitzer Space Telescope.

A totally different method employs the phenomenon of gravitational lensing, mentioned in the early chapters of this book. The light that we receive from a distant star can be brightened if a massive object (such as another star) happens to pass very close to the line of sight between us and the distant star. What we observe is a characteristic gradual brightening, followed by an identically shaped decline. The entire process can take weeks, depending on the alignment. The technique is to observe a patch of sky that is densely populated with millions of stars again and again, each time comparing the images with each other using a powerful computer. We look for small but significant changes in the brightness of any one of the stars. The shape of the 'light curve' tells us immediately whether or not this event is a case of gravitational lensing. What is especially interesting is whether there are secondary, narrower features superimposed on this light curve. These are due to gravitational lensing by the planets in orbit around the intervening star. This has been observed several times, and gives further and independent clues as to the incidence and characteristics of extrasolar planetary systems.

Finally, we get back to the most straightforward approach – just look for images of the planets. Amazingly, this has now been done a few times, using new technologies, infrared wavelengths, and the biggest telescopes we have. Rather than 'just staring at the spotlight', we have learned how to block the spotlight from our view. This requires perfect control of the images, which is most easily done in space. But now, thanks to 'laser guide stars', highly sophisticated computer control of big mirrors, and observing in the infrared, we can almost completely eliminate the disturbing influence of the Earth's atmosphere, making possible the use of the giant telescopes we have on the ground for this purpose. The choice of star is also important; for example, the presence of a thin dusty disk is a promising sign, as such disks are indicators of possible planetary systems. Several direct detections of extrasolar planets have now been achieved, although they are still just points of light to us. In one spectacular image three planets are seen, at various distances around their star. The HST, observing above the disturbing atmosphere, has also been used; it has succeeded in obtaining a direct image of a planet that is a billion times fainter than its parent star. Such observations are top priority scientific objectives for the next generation of giant telescopes and space missions, and will undoubtedly be a major endeavour in the years to come.

We are also now able to study the very formation process for stars and their planets, and will soon be able to detect extrasolar planets that are just being born. We know of many stars surrounded by dusty disks – the very protostellar disks that were involved in star formation and can still be seen as remnant disks around newly formed stars. These are also called protoplanetary disks, because planets are being or soon will be formed within them. These disks can be opaque at optical wavelengths due to their high dust content, but they are transparent at millimetre wavelengths. The Atacama Large Millimetre and Submillimetre Array (ALMA) presently being built in Chile by a joint European-American-Asian collaboration will make it possible to see planets in the process of formation. Even now the statistics of stars with such disks give important information about the incidence of extrasolar planetary systems in our galaxy.

The results from these diverse methods are rapidly enabling us to fill in the story of extrasolar planets. It now seems likely that most stars have planetary systems, and that these systems often contain more than one planet. In that case there may be even more planets in our galaxy than the hundred billion stars. And then there are all the other galaxies in the universe – it would be surprising if many of them didn't also contain planets (indeed, the discovery of a planet from another galaxy has recently been announced). We are starting to learn much more about how planetary systems form and evolve. The initial 'selection effect', in which we found mostly small-orbit 'hot Jupiters', is being corrected as we find more and more lower-mass planets with much larger orbits. A system containing five (possibly seven) planets has been found, and one of those planets has a mass of just 1.4 Earth masses; the Kepler mission has recently found a tightly packed system with six planets. We are starting to understand the 'family' of extrasolar planets. It is an important, rich and rapidly developing field of research.

Step (2) in the search for life is to find planets that may be capable of supporting life. Our own planet Earth looks like a 'pale blue dot' when seen from a great distance, so we are ideally looking for a pale blue dot near a distant star. It should be neither too close to nor too far from its parent star. This is the focus of current extrasolar planet searches. Our search techniques for such distant objects are limited to spectroscopy and imaging, and these largely limit us to searching for life on the surface of the planet (which in any case is probably the most likely situation for the existence of a civilization). And if we concentrate on life that uses water as the liquid medium, we also require surface liquid water.

With these criteria, we can now be a bit more specific about the so-called habitable zone around any given star. The planet cannot be so hot that the water would all evaporate and be lost into space, nor so cold that it would freeze. The habitable zone in our own solar system extends from about 0.8 to 1.7 times the orbit of Earth. This is a fairly wide region in the inner solar system, but the only planets in it are Earth and Mars. Other factors play a role. If a planet is too small it cannot maintain a thick atmosphere, important for greenhouse warming; Mars has only one-tenth the mass of Earth. Venus too may have been a possibility, but it's a bit too close to the Sun and suffered runaway greenhouse heating.

The habitable zone for a given star obviously depends on the luminosity of the star. The stellar luminosity changes with time, and this causes the habitable zone to change too. The properties of the planet itself can be important. Volcanism can help sustain atmospheric gases, plate tectonics can play a role through effects on the carbon dioxide cycle, and the presence of a magnetic field can shield the planetary atmosphere from being stripped away by the stellar wind.

Life may still be possible outside of the traditional habitable zone. We know from the extremophiles here on Earth that life can exist in glaciers, superheated vents, deep in the ocean, and in rock kilometres below the surface. For them, subsurface groundwater, and certainly subsurface oceans such as may exist on Jupiter's moon Europa, would seem like paradise. Water is not uncommon in the outer regions of stellar systems, nor, probably, are moons like Europa around far-out Jupiter-mass planets. In that case, the water can be kept in the liquid phase by internal heating or tidal heating of the moon, both of which are independent of the distance from the star. Another possibility could be Europa-like moons around Jupiter-like planets orbiting brown dwarfs, which are not massive enough to become nuclear-burning stars. Still another is Earth-like planets which have maintained thick atmospheres even after being ejected into interstellar space. These could conceivably have surface life. Of course there also remains the possibility of life that is not carbon–water based, and for these a whole range of other scenarios can be envisaged. So life may not be restricted to the traditional habitable zone, although that is still the most likely case, and undoubtedly the easiest to find.

There have been speculations (the 'rare Earth hypothesis') that the Earth may be exceptional in its capability to harbour life due to a number of chance coincidences, such as the presence of the Moon, which stabilizes changes in the tilt of Earth's rotation. However these speculations are not compelling, and it is widely expected that, as the extrasolar planet census becomes more complete, many Earth-mass planets capable of harbouring life will be found in the habitable zones of stars.

Once we find Earth-size planets in the habitable zone of their parent stars, how do we go on to step (3) and find indirect evidence for life on those planets? Simply observing their brightness as a function of time may give clues about whether there are continents and oceans, whether there are clouds that move with time and whether there is snow or ice that changes with the seasons. Ultimately we may have actual images of the planets. Spectroscopy could give us an indication of the temperature of the planet and the broad chemistry of its atmosphere as seen both in emission (illuminated by the parent star) and in absorption (against the light of the parent star). Already water, methane, oxygen and carbon have been detected in the atmospheres of Jupiter-mass extrasolar planets, showing that such observations are possible even with present-day technology. If life-bearing planets are like Earth, youthful atmospheres might be rich in methane (expelled by early micro-organisms) and mature atmospheres should be rich in oxygen (due to photosynthesis). There may be many such clues in the atmospheres of these planets. Observations from a distant space-craft have been made of Earth itself, to determine what telltale signs there are of life as seen from afar. The atmosphere has much more oxygen, methane and nitrous oxide than that of a lifeless planet. Water was detected in its various forms (gaseous, liquid and solid), and there was even evidence of chlorophyll. One would look for a similar atmospheric chemistry that is incompatible with nonbio-logical processes.

Extraterrestrial Intelligence?

Perhaps we can just leapfrog over the various steps above (particu-larly step 4, the actual detection of life in these distant systems), and go directly for the Big One – *intelligent* life in the universe.

What made this even remotely thinkable over the past cen-tury was the advent of our own radio technology. Until then, our 'eyes on the universe' were limited to the narrow visible (optical) part of the electromagnetic spectrum. But over the past century we have developed the ability to observe the universe over the entire electromagnetic spectrum. The broad radio band was the first of the new 'windows' to be developed, and it still remains one of the most promising for the search for extraterrestrial intelligence. Radio waves can pass unimpeded through the vast interstellar dust clouds that block our view at visible wavelengths. That means that we can detect radio waves from the murky depths of our galaxy, as well as from the distant universe, if we have big enough radio telescopes. Another important advantage is a unique, strong and narrow spectral feature that is emitted only by atomic hydrogen, the most common element in the universe. This is the famous '21-cm line', and it is located in the middle of the radio band. Any intelligent civilization in the universe would certainly know about the presence and unique significance of this line, and may well use it as a beacon to announce its presence and contact other advanced civilizations. Stars are not strong radio emitters, so they would not drown out radio signals being sent from their planets.

Possibilities of exterrestrial emission signals include the normal everyday buzz of radio and television leakage from the alien planet ('eavesdropping'). Stronger signals would come from radar, or similar high-power facilities dedicated to space communications with satellites and beyond. A third category might be strong beacons specifically made to contact newcomers (like ourselves) to the community of galactic societies, designed to be easily detectable and decoded.

In the early 1900s Guglielmo Marconi made the first wireless radio transmission across the Atlantic. It was quickly realized that such communications would in principle also be possible between different worlds in our galaxy and even beyond. In 1924 an attempt was made to 'listen in' on any possible radio signals emanating from Mars, with a negative result. After further important developments in radio technology, especially during and after World War II, Giuseppe Cocconi and Philip Morrison revisited the question of extraterrestrial communication, and suggested that radio astronomers look for possible signals from other worlds around nearby stars, preferably near the wavelength of 21 cm. As they said in their paper, "The possibility of success is difficult to estimate, but if we never search, the chance of success is zero". As we do now have the technology available to us, it seems reasonable to make at least whatever searches we can.

Frank Drake, a young radio astronomer at the U.S. National Radio Astronomy Observatory (NRAO), independently came up with the same idea, and in 1960 quickly initiated what he whimsically called Project Ozma (after a favourite childhood movie, *The Wizard of Oz*) using a new 85-ft diameter radio telescope. He estimated that signals from as far away as 12 light-years would be detectable. Although the results of this 2-month project were negative, it greatly inspired other such efforts over the following years.

A (now) amusing footnote in the story played out over a period of a few months in 1966, when the first pulsars were discovered. In the then secrecy-shrouded Cavendish Laboratory at Cambridge University, the four known sources of the pulses were nicknamed LGM-1, LGM-2, LGM-3, and LGM-4, LGM standing for 'Little Green Men'. After all, the pulsars were located throughout interstellar space and were incredibly precise in their timing – they could conceivably have been part of some vast interstellar navigation and communication system. But we now know that they are just garden-variety neutron stars.

Undoubtedly the most determined and sophisticated efforts made to date have been those of the SETI (Search for Extraterrestrial Intelligence) Institute, established in 1984. Project Phoenix, the most comprehensive and sensitive SETI programme so far conducted, targeted some 800 stars to a distance of about 240 light-years, from 1995 to 2004. It would have been sensitive to transmitters with power similar to that of Earth-based radars. To date, no detections have been made. The SETI searches use the most powerful radio telescopes in the world, such as the 1,000-ft Arecibo dish in Puerto Rico and now the Allen Telescope Array, a novel and extremely powerful array of antennas in California. The computing power alone is astounding: thanks to the rapid pace of electronic development, SETI's speed doubles every 18 months ('Moore's Law' in electronics). It is now a billion times faster than Drake's original observations, capable of searching for non-random signals in hundreds of millions of channels simultaneously. The Allen Telescope Array will be upgraded to 350 antennas from the present 42, with a far wider bandwidth. Then, perhaps by 2030, SETI may be able to check for signals in the directions of millions of star systems.

Is radio the only possibility? The optical band may be a competitor after all: we can look for flashes of light from nearby stars. A pulsed laser operating on a planet orbiting a star could momentarily outshine the star by a factor of a thousand or so, providing an easily recognizable signal. For very long distances, perhaps some combination of optical and infrared would be optimal (the infrared for cutting through the dust). Other alternatives have been proposed – gamma rays, neutrinos, gravitational waves, entangled particles, to name but a few. Optical searches are being conducted right now.

Can we crudely estimate the likelihood of success? In 1960 Frank Drake wrote a simple but now famous equation (the 'Drake equation') to estimate how many contactable civilizations there might be:

$$N = R_* \times f_p \times n_e \times f_l \times f_i \times f_c \times L.$$

This was meant to be more of a guide to understanding the variables involved, and their relative uncertainties, than an equation to give a quantitative answer.

R_* is the number of stars born in our galaxy every year that could host life-bearing planets. Very roughly, as our galaxy contains some 100 billion stars, the oldest of which are about 13 billion years old, R_* could be over eight per year (if all the stars could host planets, which is unlikely). Of course there is quite a variety of stars in our galaxy, with a variety of lifespans. We are most interested in stars similar to our Sun that exist nearby, in the solar neighbourhood of the galaxy.

The fraction of such stars that actually have planets, f_p, is becoming one of the better-known terms, thanks to the extrasolar planet discoveries described above. At present it is estimated to be somewhere in the range of 50–75%, and our knowledge of this number will rapidly improve.

Our knowledge of the number of planets per solar system that have the optimal conditions for life, n_e, will also improve rapidly over the coming years, as the families of planets around stars become better known. Do we only include planets like ours that are in the standard 'habitable zone'? Presumably we should also include planets and moons that may be able to host extremophiles. In total there may be a few relevant bodies in a typical star system.

f_l is the fraction of those planets that actually *have* life. At the moment we know of only one such planet: ours. As we don't know how life formed on Earth, we don't know how easy or difficult it would be in a typical case. The production of artificial life in the lab may give us a clue. This term is generally given the value 0.1 – but that could be off by orders of magnitude.

The fraction of planets with life that eventually produce *intelligent* life, f_i, is of course totally unknown. One view is that the development of intelligence may have been a fluke, in which case this factor would be very low. On the other hand, for those who regard intelligent life as the inevitable long-term product of evolution, this factor could be quite high, near unity. Intelligence is of course not sufficient by itself; one also needs (in some form) the other assets we have: free hands with opposable thumbs, the functional ability for communication, etc.

How likely, then, is the development of communication technology by an intelligent and capable civilization? Given adequate intelligence, language and enough time, this seems nearly inevitable, if our civilization is anything to go by. The desire to learn and develop seems insatiable. A value near unity is usually assigned to f_c.

The average lifetime of such civilizations, L, is totally unknown, and depends on many possibilities. A civilization may destroy itself, for example through nuclear holocaust or biological warfare (but remember, the destruction would have to be 100%). It may be destroyed by any of a number of catastrophies – a major eruption on the planet, a hit by a large meteorite – but many of these are rare events, and not necessarily 100% destructive. On the positive side, the civilization may quickly arrange to minimize such dangers, such as migrating on the planet as required by environmental changes, or colonizing other bodies in the stellar system so that the same disaster doesn't destroy both (or all). The value chosen by Drake for L was 10,000 years – perhaps affected by his having lived at that time under the threat of nuclear oblivion.

Drake's original guesses put N = 10, but, as pointed out above, the purpose of the equation was just to give some idea of the variables involved. You can make N almost anything you like depending on whether you are optimistic or pessimistic, and of course the value of N says nothing whatsoever about the likelihood of our being able to make contact with them.

A long lifetime of an advanced civilization with communication abilities may actually *not* be beneficial when it comes to our communicating with them. Given that we are newcomers to the ranks of 'intelligent life', and the vast timescales available, we will certainly be amongst the very least developed in this category. We have had advanced communication technologies for only the past century, miniscule compared to the billions of years that have been available to possible intelligent extraterrestrials who may have developed much earlier than we did. Of course, an implication of this is that their technologies may be so much more advanced than ours as to be totally unrecognisable to us.

The age of our galaxy completely dwarfs the age of our civilization. The oldest stars in our galaxy are over 13 billion years old, and our own Sun is 4.6 billion years old. In contrast, human civilization has been around for only 10,000 years or so, and we have had radio communication for only the past hundred years. A hundred years is nothing compared to the ages of the stars. It is highly unlikely that another society has exactly the same age as ours. It would most likely be younger or older by tens of millions of years at least, if not billions. If it were younger, it would undoubtedly not have the technology to communicate with us. If it were older, it would undoubtedly have moved on to far more advanced means of communication (if it had not destroyed itself first!). We would be in the position of jungle dwellers who communicate with drums, totally unaware that there exists a vast global network of radio communication whose waves pass right through them. Indeed, one way of finding them (indirectly) may be to find their artefacts, inanimate *or* animate. They may have left countless generations of artificially produced 'descendents' scattered through the galaxy.

That leads us to the 'Fermi Paradox'. During a lunchtime discussion at the Los Alamos National Laboratory in 1950 concerning the possible presence of extraterrestrial intelligence, Enrico Fermi posed the apparently simple question, "Where is everybody". Innocent as it sounded, this question has many ramifications. Ancient civilizations could have colonized the galaxy, either themselves or through artificial descendents. Why haven't they been here? One possibility is that we are indeed alone in our galaxy. Another is that ancient civilizations have existed, but have not colonized the galaxy for one reason or another, including their own extinction or perhaps a lack of motivation. As they may be billions of years more advanced than we are, they may exist in ways we can hardly imagine. They may have become uninterested in planets around stars, perhaps preferring habitats around black holes where the supply of energy can be far more plentiful. The possibilities boggle the mind.

In any case, we have come a long way since 1961, and are well on our way to having reasonable estimates of the first three probabilities in the Drake equation (the relevant stellar birth rate, the fraction of such stars with planets, and the fraction of those that may be habitable). The uncertainties in the others are extremely large, but then so is the number of stars in our galaxy. Whatever the prospects, the implications of a detection would obviously be huge.

Throughout this section we have focussed on the possibility of our detecting them. But what about them detecting us? While SETI is just listening, we have (inadvertently) been broadcasting too. Military radar signals as well as our favourite radio and TV comedies since the 1950s and before have produced a 'radio bubble' which is expanding outwards at the speed of light and is now well over 50 light-years in radius. Hundreds of stars lie within that volume, and our first signals will by now have swept past all of them. Their airwaves will be constantly full of our radio and TV programmes. So if there are any extraterrestrials near those stars who have tuned in with sensitive receivers, they'll already know about us.

18. What Is the Future of the Universe?

The universe is nowhere near an end. Our Sun itself, now 4.6 billion years old, has billions of years still to go. That should give us some breathing time.

The Sun will remain much as it is for the next few billion years. During that time, however, it will be gradually heating up, and this will eventually have a dramatic effect on the Earth. After 3–4 billion years the Earth will experience runaway greenhouse heating, ultimately causing the oceans to evaporate. Some seven billion years from now the Sun will start to develop into a red giant star, eventually becoming over a hundred times larger and a thousand times brighter than it is at present. The Earth's surface will be over 1,000 K. Subsequently the Sun will be subject to a series of convulsions, during which stellar winds will carry away significant amounts of mass. In a final burst eight billion years from now the Sun will eject its outer layers as a spectacular planetary nebula reaching beyond the outer regions of the solar system. As the nebula gradually dissipates, the Sun's core will be left as a small white dwarf star, which will cool and fade away into the distant future. The Earth will be charred and cold. But well before that time the Earth's intelligent occupants will hopefully have found another place to live.

The future of the Sun is not at all speculative. We know what type of star it is, and we have observed and studied a large number of such stars. There is no doubt that its future evolution will be essentially as described above.

But what about the bigger picture – the fate of the entire universe? Following the frenzied days of the reionization and quasar epochs, the universe has been calming down. The star formation rate of the universe has decreased by a factor of a hundred over the last 8 billion years. This is because there is less gaseous fuel left in and between the galaxies to feed star formation, and because the motions of everything – galaxies, stars, interstellar clouds – have settled down and become more ordered, so there is less possibility for interactions. How this will play out over the long run depends on the large scale properties of the universe; if we understand them as well as we presently think we do, then an extrapolation into the future of the universe is also not very speculative.

As discussed in Chap. 3, observations since the late 1990s have shown that the rate of expansion of the universe is actually *accelerating*. The dominant component of the mass-energy of the universe is some form of dark energy, very similar if not identical to Einstein's cosmological constant. In this case, our universe will continue to expand, at an accelerating pace, probably forever. The universe will become emptier and emptier as this expansion proceeds. More and more of the presently visible universe will shift beyond our view, and no longer be observable by us because of the finite speed of light.

It will not come to an end abruptly. It will just gradually fade away. There will still be hydrogen-burning stars into the distant future, but there will be fewer and fewer of them. Gravitationally bound structures as large as clusters of galaxies will stay together, but will be separated by increasing distances as the universe continues to expand. The matter in the universe will be increasingly locked up in the compact remnants of deceased stars. While a significant fraction of the mass of stars is expelled back into the interstellar medium, much of it stays in the collapsed stellar cores – white dwarfs, neutron stars, and black holes. The white dwarfs and neutron stars gradually radiate themselves away through various processes. Even black holes evaporate, as Steven Hawking showed, finally exploding when the process comes to an end. Eventually there will not even be any stable atoms.

The concept of entropy encompasses this entire story. Entropy is a measure of the amount of disorder in a system. The more the disorder, the greater the entropy. The natural tendency is for entropy (disorder) to increase. Stars are systems of low entropy, because they are exceptional concentrations of the gas that would otherwise be uniformly spread throughout space. We too are low entropy systems – very unlikely combinations of atoms and molecules. The future of the universe will involve the dissipation of these low entropy systems, gradually increasing the overall entropy of the entire universe. The end result will be a completely uniform system, containing no region that is any more or less likely than any other. This is sometimes referred to as the 'heat death of the universe'.

Perhaps intelligent beings will be able to maintain their own low entropy states for some time, and organize objects around them to their advantage. The Dyson sphere comes to mind. This is an artificial sphere built around a star to maximize usage of that star's emitted energy. However the sphere itself would still radiate energy at infrared wavelengths, and so would not be any more eternal than the star inside. Another science fiction concept was Hoyle's Black Cloud, which possessed intelligence (albeit a very low-temperature kind), but even that would eventually dissipate. Nevertheless, scientific discovery may still hold surprises for us, in this as in so many other areas.

In any case, our present knowledge indicates that the long-term future of the universe will be one of continuing expansion, ever-increasing entropy and ever-decreasing temperature. But this future is very long indeed: trillions upon trillions of years.

19. Why Should We Be Able to Understand the Universe At All?

Einstein once said "The most incomprehensible thing about the universe is that it is comprehensible". Why should we be able to understand anything at all about the universe? Hardly the sort of achievement one would expect from mammals emerging from the savannahs of Africa. Certainly our brains did increase threefold in size over the last few million years, and our brainpower is far greater than that of any other creature on Earth. Useful for inventing crude weapons and hunting and living in large social groups – but being able to understand inflationary cosmology in the very early universe?? If this was part of the adaptation to the lives we lived a million years ago, then it would seem that we considerably over-shot our target.

What is Knowledge?

As is the case with other topics in this book, the word knowledge is still not precisely defined, and there remain arguments over its exact meaning. However, the words knowledge and science have been closely associated with each other at least since the time of Aristotle. The word science itself comes from *scientia* in Latin, which means knowledge. In the early 1800s, the term natural philosophy was gradually replaced by the word science to refer to studies of the natural and physical world. Much of what the ancient Greeks called philosophy is now considered science – the search for truths about the natural and physical world.

Certainly over the years an understanding has developed of what is known as the scientific method, and its role in establishing knowledge about the natural and physical world. Actually, the scientific method seems so obvious that it shouldn't need a name (although the term 'common sense' is inappropriate, as scientific domains such as particle physics and cosmology are far removed from our everyday worlds of common sense). The scientific method fundamentally involves establishing hypotheses and testing them experimentally or observationally. If successful, the hypothesis becomes elevated to a theory, which may eventually become considered a 'fact' (i.e. knowledge established beyond reasonable doubt). However, progress in science is rarely so neat and tidy. It includes discovery, experimentation, observation, serendipity, inductive and deductive reasoning, imagination, curiosity, in any chronological order – anything that will lead to the ultimate goal of establishing verified knowledge.

Scientific knowledge ranges from straightforward descriptions of the real world as normally perceived (distant continents, craters on the far side of the Moon, new galaxies), to abstract theories that encompass a wide range of observations and experiments in workable scenarios that are capable of making successful predictions.

Why Do We Want to Know?

Why do we want to study the world and the greater universe around us at all? Why not just stay in our cozy little cave, and ignore the world outside?

There are at least three significant reasons, each of a fundamentally different character. Most immediate is that knowledge of our world can improve our daily lives in many practical ways. An understanding of the plants and animals around us has given us more and better food. Knowledge of metals and fire has helped us live more secure and comfortable lives. An understanding of the weather at some level has been vital. Knowledge of the patterns and movements of the stars and planets in the sky enabled us to navigate and tell time. The universe is full of bizarre objects in extreme conditions, and an understanding of these improves our knowledge of physics here on Earth, leading to a myriad of practical applications. For example, the first nuclear reactions ever known were those in the Sun.

The two other reasons have to do with culture in the widest sense. Human curiosity drives us to know how the world works, regardless of possible applications, from life on Earth to the objects in the universe to the most fundamental properties of matter. For example, knowing what is out there and how it works is the domain of astronomy and astrophysics. We want to know all we can about the contents and workings of the universe. In doing this, we apply the laws of science we know here on Earth to understand distant objects.

The most profound goal is to understand as much as we can about the ultimate origins of everything – the entire universe, the objects in it, life such as that found here on Earth, and where we come from ourselves. Such studies comprise the purest scientific endeavour one can imagine. The objective is just knowledge, for its own sake. This endeavour is an essential part of what makes us human.

Are There (Any) Limits to Scientific Knowledge?

Can we know everything? Can we have absolutely complete and perfect knowledge? Even in principle? What would that mean? Ultimately it raises the question of whether we can ever truly know the underlying world beyond our sensory perceptions – and whether we would ever realize it if we did.

Going to a somewhat less fundamental level, could we ever have complete knowledge in the sense that we could answer every question and be able to predict everything? Is the potential world of knowledge finite? Is there an end to science?

In the case of the physical sciences, does knowledge result only in formulae that make it possible to predict the future, or is it something more fundamental? Is it ultimately only an 'engineering tool', useful for predictions? Are many such theories that encapsulate our knowledge non-unique, in the sense that we could always invent another scenario, involving different concepts, that equally well describes the world we experience?

And even then – is knowledge in physics always an approximation? Unlike pure mathematics, whose truths in a Platonic sense are supposed to be perfect and eternal, physics relies on observation, experiment and measurement, and there will always be uncertainties, however small. Can physical science ever be anything more that an approximation to the truth?

There are simple ways in which our knowledge of science could never be complete, such as the possibility that we live in a world in which the laws of physics are constantly changing – perhaps rapidly and erratically. We may simply never be able to keep up, let alone come to a complete understanding. We do, however, have strong upper limits on any variation in the 'constants' of physics with either time or space. These limits come from studies as diverse as radioactive decay in a 'natural nuclear reactor' deep in a mine in Gabon, West Africa, and astronomical observations of the ratios of physical 'constants' going back 10 billion years. The limits are very tight indeed: less than a part in a million billion (10^{-15}) per year. Of course such observations do not exclude some variation over time, and all bets are off when we consider 'other universes', if they exist.

What other possible limitations on science can we think of? The universe is pretty big, and contains quite a few particles and perhaps quite a few different and complex forms of life on billions of planets. This is a quantitative issue that could conceivably be overcome by a complete understanding of the principles (a Theory of Everything) rather than recording every object. But there are much more fundamental limits. We can't know what is inside a black hole, even in principle, because no light and no information can escape its clutches; we can't see inside its 'event horizon'. We also can't know about regions in our universe that are so far away that light from those regions has not had time to reach us; our observability of the universe is limited to our 'light cone'. We can't observe other universes, even though they may exist in vast numbers; they may have very different physical laws from ours, but we'll never know. The uncertainty principle imposes an absolute limit on our ability to predict events on subatomic scales. It looks like we're pretty hemmed in.

To make matters even worse, according to Karl Popper's definition of a scientific theory, which is widely accepted, we can never have complete knowledge (or at least we wouldn't know it if we did). In this view a scientific theory is one that is falsifiable, i.e. one capable of making predictions that can be tested by experiment and rejected if false. A theory can only be falsified – it can never be *proven* to be true, no matter how many experiments have given positive results.

These are all pretty daunting obstacles. But, for a few rays of hope, consider the following. Black holes are actually not completely black; they do eventually evaporate (albeit after a very long time in the case of massive black holes). Our light cone in the universe is expanding, so we see further as time goes on (although the recently discovered acceleration of the expansion of the universe means that distant regions are receding from us at an ever-increasing rate). There may be no 'other' universes to know about. And the uncertainty principle may face an uncertain future, as new developments might give rise to modifications in quantum theory itself. It is possible, of course, that we retreat somewhat and limit our ambitions for complete knowledge to the nearest (say) five billion light-years. On the other hand, new capabilities provided by the science of the future may render the above obstacles obsolete. We just don't know. Our assessment of the future prospects of science will undoubtedly evolve over time, given that science itself is moving at an incredibly rapid pace.

Indeed, through this rapid growth in science we have become accustomed, even addicted, to progress. But is there always progress? It would certainly be upsetting for most of us to be transplanted back into the dark ages, when people lived in the shadows of gigantic structures built a thousand years previously by the Romans, and yet whose own ambitions were mere survival. They would have been amazed by stories of the huge Library of Alexandria, which was established in the third century BC to collect all the knowledge of the world. And they would have marvelled at the Antikythera mechanism, an ancient computing device containing more than 30 gears, built in the second century BC; it predicted astronomical positions and eclipses, and its technology was only surpassed some 1,500 years later. Nowadays we are so accustomed to rapid progress in science and technology that it seems unthinkable that this progress could cease, and that we could actually go backwards.

Scientists can react in different ways to progress in science. Consider a grand scientific edifice that has been painstakingly constructed and is hugely successful in that it explains many things and is capable of many accurate predictions. Suppose that a new observation or experimental result now comes along and, in a stroke, destroys this edifice. In response to this, some scientists react with horror, and others with joy. The horrified ones are shocked that such a carefully constructed and successful theory should have been destroyed. The joyful ones are delighted that, in contrast to the many experiments that have merely confirmed the theory, finally we've learned something. In both cases they want science to advance, but they see the development from opposite points of view. (Of course the theory has not actually been destroyed, it will just have to be modified to accommodate the new results; this is how science makes progress.)

Would we like to know everything, if we could? Interestingly, people react in different ways also to this question. Scientists are keen on the scientific endeavour, but not necessarily so keen on the possibility that someday it could conceivably be completed. In fact, many scientists believe that there are no limits to science – that it will continue forever, without end. And they are remarkably hostile to the idea that there could be an end to science. Why? Surely it isn't just because they would then be out of a job, or might get bored. Is it that they cannot imagine that the scientific world could be finite? Yet no one would say that we have to re-discover the west coast of South America (even though it has recently moved). Once that's done, it's done, period. Nor would they claim that we have to re-discover the galaxy NGC 253. Personally I don't see how we can claim that the world of science is either finite or infinite. We just don't know. But I would like to think that every step in the scientific process (if done correctly) is done forever – that we are making progress, rather than just running on an eternal treadmill. I also believe that there is some kind of ultimate reality underlying our studies – but of course this has been the subject of endless philosophical discussion.

To conclude this section, let us just note that in the extreme hypothetical case in which we make contact with an advanced extraterrestrial civilization and ask them to answer all our questions about life, the universe and everything, many people (including many scientists) would rather not have the answers given to them on a plate – they would rather find out for themselves.

Is Mathematics A Limitation?

Mathematics underlies much of physical science, so it is understandable that we should be concerned about its reliability. If mathematics underpins our physical understanding of the universe, then any limitations of mathematics would have implications for the completeness of science.

Was mathematics discovered or invented? The early Babylonians and Egyptians used mathematics as a tool well before the Greeks, but it was the Greeks who took mathematics to entirely new heights. Rigorous proofs were essential, not just whether the mathematics worked in practice. Plato raised mathematics to an ideal, pure, abstract ethereal plane in which its truths were eternal and perfect. As such, mathematics would not suffer the indignities of the real, imperfect and uncertain world we live in. The 'truths' of mathematics so envisaged had to be 'discovered', in contrast to the messier inventions of humans. According to Plato, this ideal realm exists, whether we know it or not. In a way the situation is similar to that of a sculpture seen half-emerged from its block of granite – was Michelangelo creating the sculpture, or was he merely uncovering the sculpture that had always been there in the rock? So – are the truths of mathematics discovered from the Platonic realms, or are they invented by humans?

A few simple examples may illuminate this topic. If there are three rabbits in a field, and then another two join them, there will be five – whether or not humans are there to observe them. The numbers exist independently of humans. The fact that mathematics developed a long time ago with no consideration whatsoever for possible applications turned out centuries later to be useful (or even critical) in science and technology is another example. Euclid's geometry and the theorems of Pythagoras are just as valid today as they were more than two millennia ago when they were written down.

Over the last two centuries new developments in mathematics have rocked Platonism. First, new (non-Euclidean) geometries arose in the nineteenth century – geometries of curved surfaces in which the angles of a triangle no longer added up to 180°. As these new geometries could be created just by making new assumptions, they came as a shock to the world of mathematics, especially to Platonism, suggesting as they did that mathematics may be a human invention after all. Second, two attempts to place mathematics on a secure foundation of axioms sufficient for all mathematics went wrong. One of these was based on set theory, which is closely related to logic. Depending on assumptions one could produce different, mutually exclusive set theories – there was no single definitive set to provide the unique basis for mathematics. The other was based on the formalism of mathematics itself. This effort was derailed by Gödel's incompleteness theorems, which prove that no significant formal system of mathematics can be constructed that is both consistent and complete – there can be no comprehensive system of mathematics. These developments were unnerving in the field of mathematics, and are of concern also for science. (It is actually a remarkable coincidence that the incompleteness theorems and the uncertainty principle appeared within a few years of each other!) But in spite of these setbacks, mathematicians have kept busy happily developing and expanding ever more areas of mathematics.

Several other lines of argument also suggest that mathematics is not 'discovered' from an ethereal world of ideals. The prime numbers arose only in Greece – not in China, Egypt or Babylon. How can humans access ideals that are not part of the physical world? How does nature know to obey the abstract rules of mathematics? Einstein asked how idealistic concepts could so perfectly fit physical reality.

Our brains presumably evolved to cope with the real world. In that case it should not be too surprising that ultimately we developed methods (mathematics) that worked for that purpose. The concepts then became idealized into abstract terms. In this view, mathematics is an abstract invention of the human brain. There are many concepts in mathematics, and only those that suit a given purpose will be selected. They may also be further refined as required. If we were not dealing with discrete units of things, we would presumably have no need for numbers. A jellyfish deep in the ocean surrounded by nothing but smooth water would not necessarily be aware of the concept of number. On the other hand, some birds can count (in small numbers), because it's useful to them. There are of course many examples of mathematical tools and systems that were invented on purpose to solve specific problems in the real world, such as Newton's calculus, which was later extended to differential equations.

Modern views coming from neuroscientists and biologists are relevant. Neuroscientists have identified specific regions of the brain that seem to play an important role in mathematical thought, and others, studying infants and individuals with brain damage, conclude that at least some elements of mathematics (basic arithmetic, and perhaps some concepts of geometry) may be innate, which brings evolution into the picture. It seems plausible that we have evolved a mathematical ability that helps us deal with the world.

But it is still remarkable that mathematics is so incredibly successful in our world. The physicist Eugene Wigner described this as "the unreasonable effectiveness of mathematics" – mathematical predictions agreeing with physical experiments to astounding accuracies approaching a part in a trillion – mathematics that not only explained existing results but correctly predicted new ones, to similar accuracies. Newton's laws of motion modelled reality in terms of pure mathematics, which ultimately produced far greater accuracies than the observations on which they were based. Mathematics plays an essential role in countless fields today.

It seems that our ability for mathematics far, far exceeds anything we needed when we emerged out of Africa. Where has this ability come from? Is it one of the results of our brains having become 'oversized' in the last few million years?

The surprising applicability of mathematics to the universe at large is at least in part due to the fortunate fact that the laws of physics are the same across the observable universe. Is our system of mathematics in some way unique? Would extraterrestrial intelligent creatures use something like the same mathematics? What about those in other universes (which may have different laws of physics and different geometries), if they exist? If mathematics resided in an ethereal Platonic world, wouldn't you expect it to be the same over the entire multiverse? But, on the other hand, you might expect that the fundamentals of mathematics would be the same everywhere even if mathematics is invented, as the concept of number is so basic to experience.

Are There Questions That Science Can Never Address?

Believe it or not, in a field which is supposed to be unfettered by preconceived notions and free for curiosity-driven research, there have been taboos in science. Subjects which were discouraged, sometimes very strongly. And in many cases we're talking about taboos that originated in or were fostered by science itself. To be fair, most of these taboos arose because it was thought that the subjects in question were not (yet) amenable to the experimental scientific method. But they sometimes had negative effects that lasted for a century, before it was realized that the subjects were acceptable after all. All of the main topics of this book have been affected at one time or another.

Perhaps the most serious of these was human nature. According to the 'blank slate' concept, humans have no differences in their innate mental abilities whatsoever. This was consistent with the 'politically correct' social norms of the time, and so became frozen in for most a century. It has only been discredited over the last few decades.

Others included the study of consciousness (behavioural psychologists thought it was beyond the reach of experimental science, and so it was taboo for nearly a century), any consideration of animal cognition (again excluded because animals can't explain themselves by talking (which is also true for human babies, who are nevertheless given the benefit of the doubt)), and the origin of life (Darwin himself excluded this from his studies, as it seemed impossible to study).

These all of which have since been removed, should give us pause about any which are imposed in the future, regardless of how well founded they may seem at the time. Even more insidious are subtle biases, which can distort the progress of science without creating obvious blockages. An example is a conservative committee rejecting funding for a potentially revolutionary novel experiment which has no guaranteed outcome, or a novel measurement of something which is 'already known'. Other examples abound. The freedoms of thought and imagination are essential to science.

So are there any subjects that science can truly never address? Anything that is a recurring part of our experience in the natural world should be accessible to the scientific method. Once-only events may not be, as there is then no way of doing subsequent experiments (although studies, considered valid, of what was once thought to be our unique universe (before the multiverse concept became popular) are instructive here).

What about questions of taste? Well, in the context of eating, it is known that bitter things are often toxic. A good case for an innate preference. Beauty? It has been said that we have a preference for symmetry in human faces, which could be an indication of innate health and therefore a good partner. A sense of fairness? As mentioned also in Chap. 14, even that may be innate. We may well question whether many other matters of taste can ever be addressed by science, but, given the record so far, it would be unwise to rule any of them out.

Needless to say, there are questions such as "Why is there something rather than nothing?" and "What is the purpose (and/or meaning) of life?" that are normally considered to lie outside the domain of science. However it is interesting to note that the question "How did *our* universe come to exist?" is currently being considered (speculatively) in the context of the 'multiverse', and that similar questions (which must have seemed equally outlandish hundreds of years ago), such as "How did the solar system (or our galaxy) form?" have now been largely answered by science. Another such question, "Why does life exist in the universe", should eventually be answerable by physics and chemistry (see Chap. 11), and the question "How did we (humans) come to exist" has already been answered by evolution. Many other well-posed questions about the material world may ultimately be addressable by science.

Another subject that comes into contact with science from time to time is the paranormal. 'Para' means 'beyond'– the paranormal is "beyond the scope of normal objective investigation or explanation" according to the Oxford English Dictionary. Examples include magic, astrology, prophecy, extrasensory perception, mental telepathy, clairvoyance, miracles, the afterlife, spirits of the dead, reincarnation, and there are many others. If they are indeed totally beyond the scope of scientific investigation or explanation, then that's that. But at least some of them are certainly not beyond the scope of science, as they make predictions about the material world, which can then be tested using standard scientific methods. Some others, which reportedly come into the awareness of people but do not have obvious effects in the outside physical world, could still someday be subject to scientific study if, as seems likely, consciousness is a function of the material brain. Although large rewards have sometimes been offered by groups or individuals for proof of paranormal claims, so far no such claims have become part of the body of rigorously established scientific knowledge.

Natural philosophy has become science. Anything that we can experience in the natural world is a potential subject for scientific study. The ultimate limits of science, if any, are unknown.

How Will Science Develop in the Future?

Where will science be 100 years from now? At the turn of the century just over a hundred years ago, it was thought that almost everything was known about physics. Only a few details to clear up, such as the aether. But then in 1900 came Planck's famous paper that started quantum mechanics, and, in 1905, Einstein's famous paper on relativity. Two huge developments, that led to atomic and nuclear physics, the hydrogen bomb, and the universe of curved space.

A hundred years ago it was thought that our galaxy was the entire universe. Now we know that the universe is vastly bigger, comprised of hundreds of billions of galaxies like our own. A hundred years ago we did not know how life worked. Now we have discovered the secrets of life and mapped the entire human genome.

Technically, of course, it's been the same story. Over the last century and a bit we have developed the 'horseless buggy' (the automobile), the telephone, radio, television, the internet, nuclear power, we have learned to fly, and we have been to the Moon. Incredible developments.

So – what will the world of science look like 100 years from now? This may well be the century of genetics. Artificial life may be created in the laboratory. Applications will abound, both beneficially in the field of medicine, and more problematically in the genetic alteration and design of life. We may, for the first time ever, be liberated from natural evolution. Through genetics we may very well take control of our own evolution, and that of countless species around us. Evolution may be far faster than ever before. By actually creating new life forms, we will have the freedom and power to create far more variety in life than has ever been done by natural evolution. This new power, of course, may be either a blessing or a curse, depending on how we handle it.

We may well have determined the nature of dark matter and dark energy, and moved on to explore the new worlds they lead to. There could be a revolution in our understanding of fundamental physics. Will a Theory of Everything still be on the agenda of fundamental physics? Will we still be uncertain as to whether we live in a multiverse? Will exotic phenomena such as entanglement have led to great new frontiers? It is safe to say that science 100 years from now will involve new discoveries that followed from discoveries that are yet to be made in this century.

And in the far distant future? Perhaps the most likely long-term outcome is that we will end up with some intermediate scenario – substantial but not total knowledge. We may have answered (most of) the major questions, but science may well continue forever, dealing increasingly with questions of less importance.

What we learn about the easily observable world (the trees, the mountains, the continents, the planets, the galaxies) will be known forever; there will never be a need to retrace our steps. The world around us will evolve, but our knowledge will keep up. We will know the genetic code forever, but the apparently inexhaustible details of microbiology and changes from evolution will never cease being a source of continuing study.

We may eventually have a 'Theory of Everything' that covers everything we know of, from the finest subatomic scales on up, but we may never be sure that it covers everything that exists, we may not necessarily know whether it's a fundamentally unique theory, and will probably not be able to prove its validity beyond the limits of observation and experimentation. Mathematics will always be essential for our physical theories, even though it is in principle limited by its own incompleteness.

According to our current understanding of physics, we will never be able to observe beyond our light cone, and the uncertainty principle will also pose a fundamental limitation. Other universes, if any, will probably be forever beyond our reach, although more clues may emerge from the properties of our own universe.

This is a conservative view, an extrapolation from the current state of science, but science always seems to have a way of surprising.

Epilogue

From scientific knowledge meticulously collected, worked out and assembled, we can follow the evolutionary trail all the way from the Big Bang to the present. The fundamental matter we're made of and the laws that govern it come from the first moments of the universe. The lightest elements have existed since the first minutes. The microwave background is the remnant of the primeval fireball, and its photons have been travelling to us for over 99.99% of the history of the universe. Carbon, the element that best defines life as we know it, is made in stars, a process that has been taking place continuously for over 13 billion years.

We can discern and study a remarkable series of events throughout the history of the universe: the fundamental processes in the first second, the creation of the first elements, the dark ages of the universe, the first stars and galaxies, reionization, the formation of our solar system, the evolution and proliferation of life on Earth, and, in the last few moments, the emergence of humans and human consciousness. We can speculate about both ends of the story – the possibility of a multiverse that might have preceded our universe, and the likely long-term future of our universe. We have discovered hundreds of planets beyond our solar system, and are making rapid progress in the search for life elsewhere. It is truly remarkable that we are now able to piece this amazing story together. We are increasingly understanding the universe and our place in it.

In following all of this, we've come across some of the 'big questions' of science. The origin of the universe(s) is still a wide-open question, in spite of the phenomenal progress of recent decades, and an answer does not seem imminent. While our universe is becoming well understood – its expansion, its evolution, the formation of the elements and the ripples that ultimately formed all the stars and galaxies – there are speculations about

P. Shaver, *Cosmic Heritage*, DOI 10.1007/978-3-642-20261-2_20,
© Springer-Verlag Berlin Heidelberg 2011

many other universes 'beyond' ours. A 'final' understanding may be a long way off.

The actual origin of life on Earth is also unknown, and may remain so forever. Nevertheless, there seem to be reasonable prospects that we may learn about it by doing it ourselves – by creating artificial life. We may develop several different ways of creating life. In that case, understanding the actual way (of the many possibilities) that life originally emerged on Earth would become just a matter of history rather than of fundamental science. However, we would still like to know whether or not life in the universe at large was inevitable, and for this we need to understand the prebiotic pathways that are ultimately favoured in the environment of a young planet. This is complex chemistry, and may take some time.

Neuroscience is developing exponentially. There is now no doubt that cognition evolved (as did everything else), and that our brains do not differ in any fundamental way from those of other mammals. However, our brains did become larger than a simple linear extrapolation would predict, and specific areas such as the prefrontal cortex became larger still. With the huge multiplier effect of the enormous number of connections in the brain, this size difference alone may suffice to explain how we have become so exceptional in our mental capabilities.

Consciousness has long been considered to be one of the great mysteries of science. But it now seems that consciousness is indeed a function of material processes in the brain, and that many of the 'neural correlates' of consciousness are presently being identified. The issue then becomes how these processes come together so seamlessly to give us such a strong sense of self and continuity. It is likely that rapid advances in neuroscience over the coming years will make consciousness less and less of a mystery.

Another one of the big questions is whether there is life beyond Earth. A giant step in this direction was the detection of the first known planets outside of our solar system. Most astronomers had little doubt that there were other planets 'out there', but until we discovered them we didn't know for sure. Now we know of hundreds. But life – that's another question.

The search goes on for life elsewhere in our solar system, and for evidence of life beyond our solar system.

Many questions have certainly been answered, while the answers to others remain frustratingly elusive. The pursuit of scientific knowledge continues as never before. We're certainly not just at the beginning, and we're probably nowhere near the end. Science is thriving, and it's an exciting time.

Further Reading

The Universe

Aczel AD (2003) *Entanglement*. Plume, New York

Barrow JD (2007) *New Theories of Everything: The Quest for Ultimate Explanation*. Oxford Univ. Press, Oxford

Barrow JD, Tipler FJ (1986) *The Anthropic Cosmological Principle*. Oxford Univ. Press, Oxford

Carr B ed. (2007) *Universe or Multiverse?* Cambridge Univ. Press, Cambridge

Clegg B (2006) *The God Effect: Quantum Entanglement, Science's Strangest Phenomenon*. St. Martin's Press, New York

Clegg B (2009) *Before the Big Bang: The Prehistory of our Universe*. St. Martin's Press, New York

Davies P (1995) *About Time: Einstein's Unfinished Revolution*. Orion Publications

Davies P (2006) *The Goldilocks Enigma: Why is the Universe just Right for Life?* Penguin/Allen Lane, London

Gasperini M (2008) *The Universe Before the Big Bang: Cosmology and String Theory*. Springer-Verlag, Berlin Heidelberg

Greene BR (1999) *The Elegant Universe: Superstrings, Hidden Dimensions, and the Quest for the Ultimate Theory*. W.W. Norton & Co., New York

Greene BR (2004) *The Fabric of the Cosmos: Space, Time and the Texture of Reality*. Alfred Knopf, US

Green BR (2011) *The Hidden Reality: Parallel Universes and the Deep Laws of the Cosmos*. Alfred Knopf, New York

Gribbin J (2009) *In Search of the Multiverse*. Allen Lane, London

Guth AH (1997) *The Inflationary Universe: The Quest for a New Theory of Cosmic Origins*. Perseus Publishing

Harwit M (1981) *Cosmic Discovery: The Search, Scope and Heritage of Astronomy*. Basic Books, New York

Hawking S (1988) *A Brief History of Time: From the Big Bang to Black Holes*. Bantam Books, New York

Hawking S (2002) *The Theory of Everything: The Origin and Fate of the Universe*. New Millennium Press, Beverley Hills, CA

Hawking S, Mlodinow L (2010) *The Grand Design: New Answers to the Ultimate Questions of Life*. Bantam Press, London

Hawking S, Penrose, R (1996) *The Nature of Space and Time.* Princeton Univ. Press, Princeton NJ

Peebles PJE, Page LA, Partridge RB, (2009) *Finding the Big Bang.* Cambridge U. Press, Cambridge

Rees M (1997) *Before the Beginning: Our Universe and Others.* Perseus Publishing, New York

Rees M (1999) *Just Six Numbers: the Deep Forces that Shape the Universe.* Weidenfeld & Nicolson, UK

Rees M (2001) *Our Cosmic Habitat.* Princeton Univ. Press, Princeton

Silk J (2005) *On the Shores of the Unknown: A Short History of the Universe.* Cambridge Univ. Press, Cambridge

Silk J (2006) *The Infinite Cosmos: Questions from the Frontiers of Cosmology.* Oxford Univ. Press, Oxford

Singh S (2004) *Big Bang: The Most Important Scientific Discovery of All Time and Why You Need to Know About It.* Fourth Estate

Smolin L (2007) *The Trouble with Physics: The Rise of String Theory, the Fall of a Science, and What Comes Next.* Mariner Books, New York

Steinhardt PJ, Turok N (2007) *Endless Universe: Beyond the Big Bang.* Doubleday, New York

Susskind L (2005) *The Cosmic Landscape: String Theory and the Illusion of Intelligent Design.* Little, Brown & Co., New York

Vilenkin A (2006) *Many Worlds in One: The Search for Other Universes.* Hill and Wang, New York

Weinberg S (1978) *The First Three Minutes: A Modern View of the Origin of the Universe.* Fontana Paperbacks

Life and Evolution

Adams F (2002) *Origins of Existence: How Life Emerged in the Universe.* The Free Press, NY

Austad SN (1997) *Why we Age: What Science is Discovering about the Body's Journey Through Life.* Wiley, New York

Bennett J, Shostak S, Jakosky B (2003) *Life in the Universe.* Addison Wesley, San Francisco, CA

Bonner JT (2009) *The Social Amoebae: The Biology of Cellular Slime Moulds.* Princeton Univ. Press, Princeton

Bonnet R-M, Woltjer L (2008) *Surviving 1,000 Centuries: Can We Do It?* Springer-Praxis Publ., Chichester, UK

Boss A (2009) *The Crowded Universe: The Search for Living Planets.* Basic Books

Cairns-Smith AG (1985) *Seven Clues to the Origin of Life.* Cambridge Univ. Press, Cambridge

Cochran G, Harpending H (2009) *The 10,000 Year Explosion: How Civilization Accelerated Human Evolution.* Basic Books, New York

Coyne JA (2009) *Why Evolution is True.* Viking Penguin, New York

Darwin C (1859) *The Origin of Species.* John Murray, London

Darwin C (1871) *The Descent of Man, and Selection in Relation to Sex.* John Murray, London

Darwin C (1872) *The Expression of the Emotions in Man and Animals.* John Murray, London

Davies P (1999) *The Origin of Life.* Penguin Books, London

Davies P (2010) *The Eerie Silence: Renewing Our Search for Alien Intelligence.* Houghton Mifflin Harcourt Publ. Co., New York

Dawkins R (1976) *The Selfish Gene.* Oxford Univ. Press, Oxford

Dawkins R (1988) *The Blind Watchmaker.* Penguin Books, London

Dawkins R (1995) *River out of Eden: A Darwinian View of Life.* Weidenfeld & Nicolson, UK

Dawkins R (1996) *Climbing Mount Improbable.* Viking

Dawkins R (2009) *The Greatest Show on Earth: The Evidence for Evolution.* Bantam Press, London

De Deuve C (1995) *Vital Dust: The Origin and Evolution of Life on Earth.* Basic Books, New York

Deamer D, Szostak JW eds (2010) *The Origins of Life.* Cold Springs Harbor Laboratory Press, New York

Dennett DC (1995) *Darwin's Dangerous Idea: Evolution and the Meanings of Life.* Allen Lane, London

Diamond J (1992) *The Third Chimpanzee: The Evolution and Future of the Human Animal.* HarperCollins, New York

Diamond J (2005) *Guns, Germs and Steel: A Short History of Everybody for the last 13,000 Years.* Vintage

Dyson F (1999) *Origins of Life.* Cambridge Univ. Press, Cambridge

Fairbanks DJ (2007) *Relics of Eden: The Powerful Evidence of Evolution in Human DNA.* Prometheus Books, Amherst NY

Fenchel T (2002) *Origin & Early Evolution of Life.* Oxford Univ. Press, Oxford

Finch CE (1994) *Longevity, Senescence and the Genome.* Univ. of Chicago Press, Chicago

Fortey R (1999) *Life: A Natural History of the first Four Billion Years of Life on Earth.* Vintage Books, New York

Fry I (2000) *The Emergence of Life on Earth: A Historical and Scientific Overview.* Rutgers Univ. Press, New Brunswick, NJ

Hayflick L (1994) *How and Why we Age.* Ballantine Books, New York

Hazen RM (2005) *Genesis: The Scientific Quest for Life's Origin.* Joseph Henry Press, Washington DC

Hölldobler B, Wilson EO (2009) *The Superorganism: The Beauty, Elegance, and Strangeness of Insect Societies.* WW Norton, New York

Holliday R (2010) *Aging: The Paradox of Life, Why we Age.* Springer, Dordrecht, NL

Impey C (2007) *The Living Cosmos: Our Search for Life in the Universe.* Random House, New York

Impey C ed (2010) *Talking about Life: Conversations on Astrobiology.* Cambridge Univ. Press, Cambridge

Jastrow R, Rampino M (2008) *Origins of Life in the Universe.* Cambridge Univ. Press, Cambridge

Kirkwood T (1999) *Time of our Lives: The Science of Human Aging.* Oxford Univ. Press, Oxford

Lane N (2005) *Power, Sex, Suicide: Mitochondria and the Meaning of Life.* Oxford Univ. Press, Oxford

Lane N (2009) *Life Ascending: The Ten Great Inventions of Evolution.* W.W. Norton, New York

Larson EJ (2004) *Evolution: The Remarkable History of a Scientific Theory.* Modern Library, New York

Luisi PL (2006) *The Emergence of Life: From Chemical Origins to Synthetic Biology.* Cambridge Univ. Press, Cambridge

Lunine JI (2005) *Astrobiology: A Multidisciplinary Approach.* Addison Wesley, San Francisco CA

MacDougall JD (1996) *A Short History of Planet Earth: Mountains, Mammals, Fire and Ice.* John Wiley & Sons, US

Maynard Smith J, Szathmáry E (1999) *The Origins of Life: From the Birth of Life to the Origins of Language.* Oxford Univ. Press, Oxford

Medina J (1996) *The Clock of Ages: Why we Age – How we Age – Winding back the clock.* Cambridge Univ. Press, Cambridge

Meyer A (2005) *Hunting the Double Helix: How DNA is Solving Puzzles of the Past.* Allen & Unwin, Sydney

Morris R (2001) *The Evolutionists: The Struggle for Darwin's Soul.* WH Freeman

Palumbi S (2002) *The Evolution Explosion: How Humans Cause Rapid Evolutionary Change.* WW Norton, New York

Pudritz et al eds (2007) *Planetary Systems and the Origins of Life.* Cambridge Univ. Press, Cambridge

Rauchfuss H (2010) *Chemical Evolution and the Origin of Life.* Springer, Berlin Heidelberg

Regis E (2008) *What is Life? Investigating the Nature of Life in the Age of Synthetic Biology.* Farrar, Straus & Giroux, New York

Ridley M (1993) *The Red Queen: Sex and the Evolution of Human Nature.* Viking, London

Ridley M (1999) *Genome: The Autobiography of a Species in 23 Chapters.* Fourth Estate, UK

Ruse M, Travis J ed (2009) *Evolution: The First Four Billion Years.* Harvard Univ. Press, Cambridge MA.

Schopf JW ed (2002) *Life's Origin: The Beginnings of Biological Evolution.* Univ. of California Press, Berkeley and Los Angeles, CA

Schrödinger E (1944) *What is Life?* Cambridge U. Press, Cambridge

Schulze-Makuch D, Irwin L (2008) *Life in the Universe: Expectations and Constraints.* Springer, Berlin Heidelberg

Shubin N (2009) *Your Inner Fish: A Journey into the 3.5-Billion-Year History of the Human Body.* Vintage Books, New York

Trotman C (2004) *The Feathered Onion: Creation of Life in the Universe.* John Wiley & Sons, Chichester UK

Tyson N, Goldsmith D (2005) *Origins: Fourteen Billion Years of Cosmic Evolution.* WW Norton, New York

Ulmschneider P (2006) *Intelligent Life in the Universe: Principles and Requirements behind its Emergence.* Springer, Berlin and Heidelberg

Vijg J (2007) *Aging of the Genome: The Dual Role of DNA in Life and Death.* Oxford Univ. Press, Oxford

Wells S (2002) *The Journey of Man: A Genetic Odyssey*. Princeton Univ. Press, Princeton NJ

Wells S (2006) *Deep Ancestry: Inside the Genographic Project*. National Geographic Society, Washington DC

Wilson EO (1992) *The Diversity of Life*. Harvard Univ. Press, Cambridge MA.

Wolpert L (2009) *How we Live and Why we Die: The Secret Lives of Cells*. Faber & Faber, London

Cognition and Consciousness

Allen JS (2009) *The Lives of the Brain: Human Evolution and the Organ of Mind*. Belknap Harvard, Cambridge MA

Allman, J (1999) *Evolving Brains*. Scientific American Library, New York

Bickerton D (2009) *Adam's Tongue: How Humans Made Language, How Language Made Humans*. Hill & Wang, New York

Blackmore S (1999) *The Meme Machine*. Oxford Univ. Press, Oxford

Blackmore S (2003) *Consciousness: An Introduction*. Hodder & Stoughton, London

Blackmore S (2005) *Conversations on Consciousness*. Oxford Univ. Press, Oxford

Blakeslee S, Ramachandran V S (1998) *Phantoms in the Brain: Human Nature and the Architecture of the Mind*. Fourth Estate

Bloom FE ed (2007) *Best of the Brain from Scientific American*. Dana Press, New York

Bray D (2009) *Wetware: A Computer in Every Living Cell*. Yale Univ. Press, New Haven CT

Byrne R (1995) *The Thinking Ape: Evolutionary Origins of Intelligence*. Oxford U. Press, Oxford

Cairns-Smith AG (1996) *Evolving the Mind: On the Nature of Matter and the Origin of Consciousness*. Cambridge Univ. Press, Cambridge

Calvin W (2004) *A Brief History of the Mind: From Apes to Intellect and Beyond*. Oxford Univ. Press, Oxford

Chalmers DJ (2010) *The Character of Consciousness*. Oxford Univ. Press, Oxford

Crick F (1995) *The Astonishing Hypothesis: The Scientific Search for the Soul*. Touchstone, New York

Damasio A (2010) *Self Comes to Mind: Constructing the Conscious Brain*. Pantheon Books, New York

Dennett DC (1991) *Consciousness Explained*. Little, Brown, USA

Dockery M, Reiss M (1999) *Behaviour*. Cambridge U. Press, Cambridge

Doidge N (2007) *The Brain that Changes Itself*. Viking Penguin, USA

Donald M (2002) *A Mind so Rare: The Evolution of Human Consciousness*. WW Norton, New York

Dowling JE (2004) *The Great Brain Debate: Nature or Nurture?* Joseph Henry Press

Edelman GM (1992) *Bright Air, Brilliant Fire: On the Matter of the Mind*. Basic Books, New York

Edelman GM (2004) *Wider than the Sky: The Phenomenal Gift of Consciousness.* Yale U. Press

Edelman GM (2006) *Second Nature: Brain Science and Human Knowledge.* Yale Univ. Press, New Haven

Edelman GM, Tononi G (2000) *A Universe of Consciousness: How Matter Becomes Imagination,* Basic Books, New York

Elman JL et al (1996) *Rethinking Innateness: A Connectionist Perspective on Development.* MIT Press, Cambridge MA.

Gadau J, Fewell J eds (2009) *Organization of Insect Societies: From Genome to Sociocomplexity.* Harvard Univ. Press, Cambridge, MA.

Gazzaniga MS (2008) *Human: The Science Behind what makes us Unique.* HarperCollins, New York

Gould JR, Gould CG (1994) *The Animal Mind.* Scientific American Library, New York

Gould JR, Gould CG (2007) *Animal Architects: Building and the Evolution of Intelligence.* Basic Books, New York

Greenfield S (2000) *The Private Life of the Brain.* Allen Lane, London

Griffin DR (1992) *Animal Minds.* Univ. of Chicago Press, Chicago

Hansell M (2007) *Built by Animals: The Natural History of Animal Architecture.* Oxford U. Press, Oxford

Hauser M (2000) *Wild Minds: What Animals Really Think.* Allen Lane, London

Ingram J (2005) *Theatre of the Mind: Raising the Curtain on Consciousness.* HarperCollins, Toronto

Kandel ER (2006) *In Search of Memory: The Emergence of a New Science of Mind.* W. W. Norton, New York

Kappeler P ed (2010) *Animal Behaviour: Evolution and Mechanisms.* Springer, Berlin Heidelberg

Koch C (2004) *The Quest for Consciousness: A Neurolobiological Approach.* Roberts & Co., Englewood CO

Laureys S, Tononi G (2009) *The Neurology of Consciousness: Cognitive Neuroscience and Neuropathology.* Academic Press, London

Lynch G, Granger R (2008) *Big Brain: The Origins and Future of Human Intelligence.* Palgrave Macmillan, New York

Macphail EM (1998) *The Evolution of Consciousness.* Oxford Univ. Press, Oxford

Passingham R (2008) *What is Special about the Human Brain?* Oxford Univ. Press, Oxford

Penrose R (1989) *The Emperor's New Mind: Concerning Computers, Minds, and the Laws of Physics.* Oxford Univ. Press, Oxford

Penrose R (1994) *Shadows of the Mind: A Search for the Missing Science of Consciousness.* Oxford Univ. Press, Oxford

Pinker S (1994) *The Language Instinct: How the Mind Creates Language.* William Morrow

Pinker S (1997) *How the Mind Works.* W. W. Norton, New York

Pinker S (2002) *The Blank Slate: The Modern Denial of Human Nature.* Allen Lane, London

Pinker S (2007) *The Stuff of Thought: Language as a Window into Human Nature.* Viking Penguin, New York

Plomin R et al (2008), *Behavioural Genetics*. Worth Publishers, New York

Pockett S, Banks W, Gallagher S (2009) *Does Consciousness Cause Behavior?* MIT Press, Cambridge MA

Ramachandran V (2003) *The Emerging Mind*. Profile Books, London

Ramachandran V (2003) *A Brief Tour of Human Consciousness*. Profile Books, London

Ramachandran V (2011) *The Tell-Tale Brain: A Neuroscientists' Quest for What Makes Us Human*. W. W. Norton & Co., New York

Reznikova Z (2007) *Animal Intelligence: From Individual to Social Cognition*. Cambridge Univ. Press, Cambridge

Ridley M (1996) *The Origins of Virtue*. Viking/Penguin, London

Ridley M (2003) *The Agile Gene: How Nature turns on Nurture*. HarperCollins, New York

Rose S ed (1998) *From Brains to Consciousness? Essays on the New Sciences of the Mind*. Princeton Univ. Press, Princeton, NJ

Rose S (2005) *The Future of the Brain: The Promise and Perils of Tomorrow's Neuroscience*. Oxford Univ. Press, New York

Rutter M (2006) *Genes and Behavior: Nature-Nurture Interplay Explained*. Blackwell Publ., Malden, MA.

Seeley TD (2010) *Honeybee Democracy*. Princeton Univ. Press, Princeton NJ

Shettleworth SJ (2010) *Cognition, Evolution, and Behavior*. Oxford Univ. Press, New York

Striedter GF (2005) *Principles of Brain Evolution*. Sinauer, Sunderland MA

Taylor, J (2009) *Not a Chimp: The Hunt to find the Genes that make us Human*. Oxford Univ. Press, Oxford

Tomasello M, Call J (1997) *Primate Cognition*. Oxford Univ. Press, New York & Oxford

Velmans M (2009) *Understanding Consciousness*. Routledge, Hove, Sussex

von der Malsburg C, Phillips WA, Singer W eds (2010) *Dynamic Coordination in the Brain: From Neurons to Mind*. MIT Press, Cambridge MA

Wasserman E, Zentall T eds (2006) *Comparative Cognition: Experimental Explorations of Animal Intelligence*. Oxford Univ. Press, New York

Wilson OE (2000) *Sociobiology: The New Synthesis*. Harvard Univ. Press, Cambridge, MA

Winston R (2002) *Human Instinct: How our Primeval Impulses Shape our Modern Lives*. Bantam Press, London

Winston R (2003) *The Human Mind: And How to Make the Most of It*. Bantam Press, London

Wright L (1997) *Twins: And What They Tell Us About Who We Are*. John Wiley & Sons, New York

Wynne CDL (2001) *Animal Cognition: The Mental Lives of Animals*. Palgrave Macmillan

Wynne CDL (2004) *Do Animals Think?* Princeton Univ. Press, Princeton NJ

Scientific Knowledge

Barrow JD (1998) *Impossibility: The Limits of Science and the Science of Limits.* Oxford Univ. Press, Oxford

Brockman J ed (2008) *Science at the Edge: Conversations with the Leading Scientific Thinkers of Today.* Union Square Press, New York

Bryson B (2003) *A Short History of Nearly Everything.* Doubleday, London

Davies P, Adams P (1996) *The Big Questions.* Penguin, Australia

Goldstein R (2006) *Incompleteness: The Proof and Paradox of Kurt Gödel.* W. W. Norton, New York

Horgan J (1996) *The End of Science: Facing the Limits of Knowledge in the Twilight of the Scientific Age.* Addison Wesley

Kuhn TS (1962) *The Structure of Scientific Revolutions.* Univ of Chicago Press, Chicago

Livio M (2009) *Is God a Mathemetician?* Simon & Schuster, New York

Maddox J (1999) *What Remains to be Discovered: Mapping the Secrets of the Universe, The Origins of Life, and the Future of the Human Race.* Touchstone, New York

Morris R (2002) *The Big Questions: Probing the Promise and Limits of Science.* Times Books, New York

Weinberg S (1992) *Dreams of a Final Theory: The Scientist's Search for the Ultimate Laws of Nature.* Pantheon, New York

Wilson EO (1998) *Consilience: The Unity of Knowledge.* Alfred A. Knopf, New York

Acknowledgements

First, of course, I owe a huge debt of gratitude to my wife Jenefer, who was with me all the way through this extraordinary journey of exploration and discovery. She was wonderfully enthusiastic about all facets of the project. She read all versions of the book, scribbling notes and comments along the way. Every night we spent hours over dinner debating the issues that came up that day. Her background in literature and medicine added fascinating and thought-provoking slants on the purely scientific issues I was pursuing. She attended some of the conferences with me, and met several of the people I had contacted. She tolerated my obsession with great dignity, and thankfully had her own interests and wonderful circle of friends so that she could escape my obsessive pursuits. Our daughter Nikki and son Adam were also very interested, read drafts and provided valuable feedback. Several friends contributed their views, in particular Christopher Abbott, George Robillard and Russell Stewart, who kindly read and commented on the semi-final draft.

I am greatly indebted to scientists in various fields who generously took the time to discuss their specialties by phone, email, or in person: Cori Bargmann (Rockfeller University, New York), Max Bennett (University of Sydney), Dorret Boomsma (VU University, Amsterdam), David le Couteur (University of Sydney), George Church (Harvard University), Nicky Clayton (Cambridge University), Paul Davies (State University of Arizona), John Ellis (CERN), Ron Ekers (CSIRO Astronomy and Space Science, Sydney), Bob Fosbury (STScI, ESO, Munich), John-Dylan Haynes (Bernstein Centre for Computational Neuroscience, Berlin), James Gould (Princeton University), Seth Grant (Cambridge University), Pattrick Haggard (University College London), Mike Hansell (University of Glasgow), Len Harrison (WEHI, Melbourne), Margo

Honeyman (WEHI, Melbourne), Gerda Horneck (DLR, Cologne), Dan Lahr (University of Massachusetts), Bruno Leibundgut (ESO, Munich), Nate Lo (University of Sydney), Ryszard Maleszka (Australian National University, Canberra), Michel Mayor (University of Geneva), Murugappan Muthukumar (University of Massachusetts), Richard Passingham (University of Oxford), Jim Peebles (Princeton University), Jean-Loup Puget (Institut d'Astrophysique Spatiale, Paris), Martin Rees (Cambridge University), George Robillard (University of Groningen, The Netherlands), Seth Shostak (SETI Institute), Joe Silk (University of Oxford), Steve Simpson (University of Sydney), Christina Smolke (Stanford University), Marla Sokolowski (University of Toronto), David Stamos (York University, Toronto), Jill Tarter (SETI Institute), Chris Tinney (University of New South Wales, Sydney) and Lo Woltjer (Obs. de Haute Provence).

I am especially grateful to the following experts who reviewed the various sections of an advanced draft of the book: Ron Ekers (scientific knowledge), Gerda Horneck (astrobiology), Bruno Leibundgut (astronomy and cosmology), Michel Mayor (extrasolar planets) and Stephen Simpson (the life sciences). Their opinions and comments were extremely helpful in finalizing the book.

Access to the latest scientific publications in all fields was facilitated through an Honorary Associateship kindly granted to me by the School of Physics, Faculty of Science, the University of Sydney.

My editor at Springer, Ramon Khanna, was very encouraging and helpful. We had some great discussions about several of the topics covered, and he made many perceptive comments that helped to improve the final version of the book.

Finally I must repeat the standard boilerplate statement that is made in so many other books: any and all mistakes in the book are mine alone.

About the Author

The author is Canadian, obtained a PhD in astrophysics at the University of Sydney in Australia, and spent most of his career as a senior scientist at the European Southern Observatory (ESO), based in Munich. He has authored or co-authored over 250 scientific papers, and edited six books on astronomy and astrophysics. His interests have ranged from our galaxy to distant quasars and the reionization of the universe. He was instrumental in the establishment of Europe's participation in the Atacama Large Millimeter/submillimeter Array (ALMA) project. A member of various international organizations and committees, he served as President of the International Astronomical Union's Division on Galaxies and the Universe. Now retired, he and his Australian wife split their time between Sydney, Toronto, and Europe, and he devotes himself to broadening his horizons in science, in the process having written this book.

Index